Orange Chapin

The Chapin Genealogy

Orange Chapin

The Chapin Genealogy

ISBN/EAN: 9783337366407

Printed in Europe, USA, Canada, Australia, Japan

Cover: Foto ©berggeist007 / pixelio.de

More available books at **www.hansebooks.com**

PROCEEDINGS

AT THE MEETING OF THE

CHAPIN FAMILY,

IN

SPRINGFIELD, MASS.,

SEPTEMBER 17, 1862.

———•••———

SPRINGFIELD:

PRINTED BY SAMUEL BOWLES AND COMPANY

1862.

PRELIMINARY RECORDS AND REMARKS.

THE CHAPIN family is one of the largest in this country. It is descended from Dea. SAMUEL CHAPIN, who settled in Springfield, Mass., in 1642—a man whose name and blood are now to be found in every one of the United States. As the proceedings recorded herein will show, it now contains some of the first minds of the nation, and has been, throughout its history, a family of intelligence, influence and virtue. In harmony with the new interest everywhere observable in genealogical and antiquarian matters, it was determined that there should be a meeting of this family. The first suggestion came from the Hon. HENRY CHAPIN of Worcester, formerly Mayor of that city, to the Hon. STEPHEN C. BEMIS, Mayor of Springfield and a possessor of the Chapin blood. The following is the letter:

JUDGE CHAPIN'S LETTER.

WORCESTER, Jan. 17th, 1862.

Hon. S. C. BEMIS—DEAR SIR:—I write you upon a matter in reference to which some of the family of "Chapins," have been exercised, and that is a meeting of the Chapin family. The natural place of such a meeting is the Valley of the Connecticut, and Rev. E. H. Chapin would naturally be the orator of the occasion. Now, for one, I should be very glad to have such a meeting held, and as I am informed that you are one of the stock, and as, from my experience, I know that Mayors are expected to know how to do everything, and to do it, I write to you and suggest that such a meeting should be held, and that you should take the initiative steps in the matter.

Please to excuse this apparent abruptness in a stranger, but I am writing in good faith, and simply say what I mean. At any rate, will you do me the favor to give me your views upon the subject?

Yours Truly,
HENRY CHAPIN.

MAYOR BEMIS' REPLY.

SPRINGFIELD, MASS., Jan 21, 1862.

Hon. HENRY CHAPIN, Worcester, Mass.—Dear Sir: Your letter of the 17th inst. is at hand. I have conversed with Chester W. Chapin, Esq., on the subject to which you refer, and he is very favorably impressed with your proposition, and suggests that you designate some early day to meet a few of the "Chapins," whom we can call together in this city and vicinity, and talk over the matter, and make arrangements for the general gathering. If you will designate some day that you will be here, I will invite some of the family to meet you at the Massasoit House in this city. I have addressed a note to Dr. Holland of this city, who will meet yourself and others at the time you may designate, for consultation. I have also addressed a note to Dr. E. H. Chapin of New York, on the subject, and shall have an answer from him soon.

Please advise me further at your convenience, and oblige

Yours,   S. C. BEMIS.

According to the suggestion of Mayor Bemis, Judge Chapin of Worcester came to Springfield, and participated in the preliminary proceedings. The general plan of the meeting was decided upon, and general and local committees were appointed to carry the plan into operation. Judge Chapin of Worcester was fixed upon by the rest of the Committee of Arrangements, as the Orator of the occasion, and Dr. J. G. HOLLAND of Springfield was appointed to pronounce a poem. The following circular was published in *The Springfield Republican* and other papers, and in note form was sent by mail to all known representatives of the family throughout the country.

CIRCULAR.

SPRINGFIELD, MASS., July 26, 1862.

To THE DESCENDANTS OF DEA. SAMUEL CHAPIN:

Through the suggestions of individuals and the operations of a self-constituted committee, it has been decided to hold a grand gathering of the descendants of Deacon SAMUEL CHAPIN, who settled in Springfield at the early date of 1642. The family has become a very numerous one, and it is believed that nearly if not quite all who bear the name in the United States, are the descendants of this man, and that their origin is traceable to the Valley of the Connecticut River.

As the old town of Springfield was the place from which the family went out, it has been deemed fitting that it be the gathering place of its members, large numbers of whom still remain upon the original soil.

A large Committee of Arrangements has been appointed, who, at a meeting held for the purpose, fixed upon the seventeenth day of September, 1862, as the time of the proposed family re-union. To this meeting you are respectfully and cordially invited; and as it is impossible for the committee to know the names of all the members of the family, scattered over the country, you are particularly requested to extend this invitation to all of the blood living in your vicinity. It is expected that the meeting will be one of public as well as social interest, and it is believed that it will do much to cherish that laudable pride, which rejoices in an excellent ancestry, and that affection which should always flow in the channels of kindred blood.

We invite you, therefore, whether you bear the name and blood of the old patriarch of the valley, or the blood without the name, to come back to the home from which you went out, and join the assembly we propose.

The public exercises of the occasion will consist of an Historical Address, by Hon. HENRY CHAPIN, of Worcester, with brief addresses by other members of the family, and music prepared for the occasion. At the close of the public exercises, the City Hall will be thrown open for a social re-union.

The Mayor's room at the City Hall will be open on the 16th of September, and on the morning of the 17th, for the registry of the names of all who wish to attend the exercises. All will be furnished with tickets on registering their names, and no person will be admitted without one.

All who accept this invitation are requested to inform the Committee on Finance and Invitation, at an early date, addressing RODERICK BURT, Springfield, Mass., that entertainment may be secured for them so far as possible; and they will please call on Mr. Burt, at his Bookstore, on their arrival in the city.

COMMITTEE OF ARRANGEMENTS.

Hon. S. C. BEMIS, Springfield, Mass.
Hon. CHESTER W. CHAPIN, "
Hon. HENRY CHAPIN, Worcester, Mass.
J. G. HOLLAND, Springfield, Mass.
GEORGE W. CHAPIN, Providence, R. I.
E. A. CHAPIN, Rutland, Vt.
MARVIN CHAPIN, Springfield, Mass.
ABIJAH W. CHAPIN, "
LYMAN CHAPIN, Albany, N. Y.
Hon. MOSES CHAPIN, Rochester, N. Y.
CALEB T. CHAPIN, Northbridge, Mass.
CHARLES F. CHAPIN, Milford, Mass.
DAVID CHAPIN, Boston, Mass.
E. H. CHAPIN, D. D., New York City.
EDWARD P. CHAPIN, Buffalo, N. Y.
D. M. CHAPIN, Ogdensburg, N. Y.
SAMUEL D. CHAPIN, Cleveland, Ohio.
M. W. CHAPIN, Hartford, Conn.
Doct. CHAS. CHAPIN, Brattleboro, Vt.
Rev. ALBERT HALE, Springfield, Ill.
ASAHEL CHAPIN, New York City.
REUBEN S. CHAPIN, New York City.
PHINEAS CHAPIN, New Hampshire.
Rev. Dr. CHAPIN, Pres. Beloit Col., Wis.
A. W. CHAPIN, Newport, Maine.
SUMNER CHAPIN, Orrington, Maine.

P. W. BARTLETT, Covington, Ky.
H. JUDSON CHAPIN, Holyoke, Mass.
GILES W. CHAPIN, Chicopee, Mass.
MOSES W. CHAPIN, "
ORANGE CHAPIN, "
MARSHAL PEASE, "
SIDNEY CHAPIN, "
LUCAS B. CHAPIN, "
PHINEAS STEDMAN, "
ETHAN S. CHAPIN, Springfield, Mass.
RODERICK BURT, "
FREDERICK H. HARRIS. "
ABEL D. CHAPIN, "
S. A. BEMIS. "
A. L. CHAPIN, "
H. ALEXANDER, JR., "
WILLIAM BLISS. "
E. BLISS VINTON, "
Col. HOMER FOOT. "
Hon. GEORGE DWIGHT, "
WILLIAM B. BRINSMADE, "
EDMUND D. CHAPIN, "
ANDREW J. CHAPIN, "
ROSWELL LEE, "
CHARLES O. CHAPIN, "
S. E. BAILEY, "

COMMITTEE ON FINANCE AND INVITATIONS.

Mayor BEMIS,	A. W. CHAPIN,	CHARLES O. CHAPIN.
MARVIN CHAPIN,	RODERICK BURT,	H. ALEXANDER, JR.

COL. HARVEY CHAPIN, *Marshal of the Day.*

In addition to the committees mentioned in the circular, were the following:

On Printing.—STEPHEN C. BEMIS, J. G. HOLLAND.
On Refreshments.—MARVIN CHAPIN, E. S. CHAPIN, A. W. CHAPIN, S. E. BAILEY, CHARLES O. CHAPIN, PHINEAS STEDMAN.
Music.—CHARLES O. CHAPIN.

The day appointed for the meeting was an exceedingly pleasant one; and the registry of the names of those present, appended to this report, will show how large the gathering was. It filled the First Congregational Church, its place of meeting, to repletion, allowing no room for citizens generally, multitudes of whom desired to be present at the exercises. The music of the occasion was given by a select choir under the charge of CHARLES O. CHAPIN of Springfield. At the close of the somewhat protracted exercises at the church, a procession was formed under the direction of Col. HARVEY CHAPIN, Marshal of the day, and, to the music of the Armory Band, marched around Court Square to the City Hall, where the remainder of the day was devoted to feasting and social enjoyment. It will be proper to record that the first portion of the procession, composed exclusively of the family, began to enter the hall before the last had emerged from the church, thus passing nearly around Court Square, from three persons to six and seven abreast.

THE PUBLIC EXERCISES.

THE exercises of the day were commenced at the First Congregational Church, at ten and one-half o'clock, by a voluntary on the organ, after which the following song was sung, by the choir:—

"Joy! joy! freedom to-day,
Care! care! drive it away;
Youth, health and vigor our senses o'erpower;
Trouble, count it for naught,
Banish, banish the thought;
Pleasure and mirth shall rule o'er this hour.
Joy to-day, joy, joy, to-day,
And care, oh drive it far away.

"Nature all her glory showing,
Azure skies and balmy air,
Equal smiles on all bestowing,
Bids each heart her bounty share.
Joy! joy!" &c.

ADDRESS OF WELCOME.

Hon. STEPHEN C. BEMIS, Mayor of Springfield, then addressed the audience, as follows:

LADIES AND GENTLEMEN,—MEMBERS OF THE CHAPIN FAMILY FROM THE EAST AND FROM THE WEST, FROM THE NORTH AND FROM THE SOUTH:—In behalf of those members of the family who reside in this valley, in behalf of the authorities and citizens of this city, I bid you a cordial welcome to Springfield, to our homes and to our firesides.

We meet on the spot where our forefathers worshipped, and on the soil rendered sacred as the depository of their dust, and consecrated by their prayers, their tears, and their trials. No sculptured marble marks the place where they rest, but they are remembered.

in the affections and in the hearts of their posterity. We may not be able to point you to improvements so vast or a soil so productive as may be found in some other portions of our country, where some of you reside, and where enterprise and energy realize a more sudden reward; we can, nevertheless, show you a city of some twenty thousand inhabitants, industrious, energetic, and frugal in their habits; we can show you a soil which richly repays for all the labor bestowed upon it: and a population in the country around us whose intelligence, and habits of sobriety and industry, are second to no other. Many of you may have no recollections of Springfield except from the traditions of the past, and in looking around you, you may well ask—

"Is this the land our fathers loved?
Is this the soil on which they moved?
Are these the graves they slumber in?"

We can show you the spot on which Samuel Chapin lived and reared his family—the old burial ground where his remains were first deposited, and the cemetery where his bones now rest. We can show where stood the old fort, built in 1660, as a protection from the Indians, and taken down in 1831; and we can show you the fields which our fathers cultivated, and the old homestead, where they lived, and where they died.

It is two hundred and twenty years since Deacon Samuel Chapin took up his abode on the banks of this river, and within one hundred rods of the place where we are now assembled. If he could rend asunder the cerements of the tomb, and appear among us to-day, and look over this vast assembly, he could not say, as did Logan, the Mingo chief, when bereft of all his kindred, that "not one drop of his blood ran in the veins of any living creature;" but rather would he say that the promises of God to the patriarchs of old had been fulfilled to him, when He said: "I will make thy seed as the stars of heaven in number, and as the sands of the sea-shore innumerable." Two hundred and twenty years! What memories rise before us as we look back upon the past! A pilgrim band, marching from the sea, through a pathless wilderness, they paused on yonder hill to admire the lovely landscape:

"Sweet fields beyond the swelling flood,
Stand dressed in living green."

Delighted with the spot, they commence a settlement. Of the

trials and dangers through which they and their descendants have passed, up to the present hour, it is not for me to speak.

We meet, my friends, when clouds hang over our country. Many hearts are desolate and sad; many door-ways are hung with the blackest mourning, and many households miss the industrious hands that once furnished them with bread. The enemies of our country are now knocking at our very doors. What shall we do? Look supinely on? No! if our country wants means, give them; if men, offer them. We know that many who bear the family name and blood are in the ranks of the Union army. One is at the head of a gallant regiment from a sister State, of whom honorable mention has been made. There are others, as officers and soldiers, who have been brave and true in the hour of trial. Our armies are now engaged in battles in which they are victorious; may they be the precursors of yet further triumphs, the defeat of the enemies of the Union, and the re-establishment of order and a love of the Constitution and the laws throughout the entire country! Let us swear anew our devotion to country and the glorious old flag, and if called to join the armies of our country, in defence of its rights, let us go not

> " Like the galley slave, at night, scourged to his dungeon,
> But sustained and soothed by an unfaltering trust."

After the close of the exercises of this day, there is no probability that we shall all meet again during the remainder of our earthly journey. God grant that we may be ready for the final hour, when we shall exchange the mortal for immortality, and may we all meet in a better and more glorious land!

I renew my welcome to you all, and trust that your visit to the old home may be pleasant and agreeable. [Applause.]

PRAYER BY REV. MR. PARSONS.

O, Thou who art the God of all the families of the earth, we invoke Thy divine presence to bless us now, and to hallow this occasion. We praise Thee, O God, our Creator, that Thou hast formed the race of man into families, and that, in Thy holy word Thou hast declared Thyself a covenant-keeper to all the families of the righteous. We praise and glorify Thy name, that thou hast left on record the descent of Thy Son, our most holy Redeemer, through the appointed channel, and hast thereby disclosed unto us that Thou art truly a God, holy and righteous, keeping Thy covenant forever, and " showing mercy unto thousands of generations of them that love

2

Thee and keep Thy commandments." We thank Thee, O God, and praise Thy name, that we are assembled here to-day to celebrate the memory of one who was known here two centuries ago, and who walked before Thee in this earthly church a faithful officer and holy man; and also preserved and transmitted to his posterity here assembled to-day, the priceless heritage of a pure faith and a holy life. We beseech Thee now, O God, as the memories of that ancestor shall here be renewed,—as these, his descendants and friends, are assembled to commemorate his memory and character, and to remember all the grace with which Thou hast led them and their ancestors from that time to this,—we do beseech Thee by Thy spirit here to abide with us, preside over our hearts, and lead us each and all to draw lessons of virtue, wisdom and holiness from those records that we shall hear, and from all that shall be said of the holy men who have gone to rest.

We thank and praise Thy name that Thou hast recorded in Thy word a promise unto those who keep Thy name and honor their fathers. Thou dost say of Jonadab, the son of Rechab, that "Jonadab, the son of Rechab, shall not want a man to stand before Thee forever," because his children kept the commandments and faith of their fathers. So do we praise Thee that Thou hast graciously fulfilled Thy covenant to the patriarch of this family, in that there has not been lacking, in the generations since his departure, a man to stand before the Lord in the church and bear his standard and be an officer of the church. We glorify Thy name for Thy mercy, for Thy loving kindness, for Thy tender care of those who have risen from him; and we beseech Thee now that Thou wilt bless these, his numerous descendants, who are here gathered. Give to them each one the fear of the Lord God. If any of them have departed from the faith of that first father, we pray Thee constrain them, upon this anniversary occasion, to renew their covenants before the Lord Jehovah, and follow the God of their fathers.

We beseech Thee, O Lord God, that Thou wilt bless this family in all its branches. May those who are not here be under Thy especial guidance, and through Thy rich mercy have that hope in Jesus Christ which shall last when this world ends. We beseech Thee, grant unto us all here, that Thy word may be precious to us; that Thy truth may be indeed in our hearts; that each one of the descendants of that holy man who walked before God in this church two centuries since may have the same faith in Thee, and through that blood which cleanseth from all sin, may have good hope of acceptance in Jesus Christ the righteous.

Let Thy spirit abide with us. Bless the families of those here assembled. Be with them, we pray Thee, and teach them how to train up those whom Thou hast given them, that they may glorify Thee in their day and generation, and be worthy of that ancestor whose name is precious and commemorated here to-day. Wilt Thou forgive us our sins. Wilt Thou grant unto us acceptance in these petitions, and Thy most holy blessing to rest upon us each one. Even now grant us the pardon of our sins, and eternal life, through Jesus Christ, Thy Son. Grant, we pray Thee, to those families who are assembled in this, Thy house, to-day, that when this life shall end, they may go to sleep with good hope in the resurrection of all the Christian just, of a reunion in heaven, to join him whose name will be on their lips to-day, and all his descendants who through Jesus Christ have found that "rest which remaineth for the people of God."

Hear us in these petitions and prayers. Grant us Thy most holy presence and blessing in all the exercises of this hour, and all the exercises of this day, and accept us through Jesus Christ, to whom, with Father and Spirit, be glory forever—Amen.

The principal address of the occasion was then delivered by Hon. HENRY CHAPIN, of Worcester.

JUDGE CHAPIN'S ADDRESS.

THIS is a meeting of the Chapin family. We have come together from different sections of the country, drawn by an elective affinity which runs in the blood. We claim as our own all of you who have happened to connect yourselves with the tribe, and for one day, at least, whatever may have been the name which you inherited from your fathers, or which, for domestic purposes, you have consented to assume, you sink your own identity, and imagine that your name is Chapin.

The part which the speaker took in getting up this meeting, was not taken with the remotest idea that he was to occupy this responsible position. Our eyes were all turned towards one whose abilities and eloquence have shed lustre upon the family name, and we doubted not that his words would thrill our hearts upon this joyous occasion. The hand of disease is upon him. Banished by the stern decree of his physician from his pulpit and his home, he wanders in foreign lands, in pursuit of the boon of health once vouchsafed to him in such rich exuberance, while we vainly lament the absence of one who entered most heartily into the plan of this gathering, and who would have gladly contributed to its successful accomplishment.

When the invitation to perform this duty reached me, it struck me with disappointment and discouragement. I felt that among the descendants of the sons and daughters who settled in Western Massachusetts, there must be numerous persons far better qualified than myself to address this large assembly, and that it was to be deplored that the committee had fixed upon a descendant of that son of the family who was so early separated from the home circle, and who neglected this fertile valley for more sterile regions nearer the great sea.

It is not easy to hold a family meeting without doing and saying many things which to outsiders may seem to savor of egotism. Without these it would hardly possess the characteristics of family confidence and sympathy. Every person knows that when we are surrounded by our own kindred and friends, the tongue becomes loosened, the spirits are set free, and unconsciously sending care and caution to the winds, we say what we think in our own way, regardless of consequences, and fearless of criticism. It is the feeling which produces this state of things, which perhaps more than any other cause gives to such a meeting its pleasure and importance. It is well to throw aside the cares, perplexities and jealousies of life, and to feel, for the moment, at least, that we are in the midst of friends whose charity for us shall cover a multitude of sins. With this feeling we come hither to-day. We have had no general meeting of our family for a very long period of time, and now when we are assembled upon our own ground, and upon our own business, we may be eloquent or not, grammatical or not, genealogical or not, egotistical or not; and kindness and good will shall sanctify all our efforts, and the attempt to make the day pass pleasantly and sociably shall be a sufficient excuse for any errors in taste or judgment.

Had the speaker been called upon to enlighten the family in his own section of the commonwealth, he might have hoped that he had gained some information as to the family history which would be interesting to them, but to come up to the old homestead to which his immediate forefathers had become strangers, and to undertake to present to the descendants of Japhet and Henry and Catherine and Sarah and Hannah anything either instructive or interesting, seemed to imply a degree of confidence in one's capacity and intelligence to which he could lay no claim. But what was to be done? The rest of the Committee had assembled without any notice to him to be present, and had quietly informed him by letter that the Rev. Dr. E. H. Chapin was sick and obliged to visit Europe, and that they had unanimously elected the speaker to deliver the address upon this occasion, and had adjourned. Although provoking, this state of things was interesting, because it might be developing a family trait. It has been said that genius is of two kinds. Genius of one kind leads a man to do a thing himself, while genius of the other kind is shown in the ability to make some one else do it. Which is the most useful and effective to its possessor, has never yet been entirely settled. From what I have seen of our family, although they are not deficient in genius of the first description, I am inclined to the belief that the latter is most fully developed in them. At any rate, it had placed the speaker in the predicament referred to, and he was obliged either to show a want of courage and dodge the meeting, or a want of prudence and make a sacrifice of himself, in order that there should be no hindrance to the contemplated family gathering. There seemed to be no escape, except by demonstrating to the family that a scion of the stock of Josiah was a coward; and this demonstration was one of the last which any sensitive man would desire to make of himself. Therefore I appear before you at the risk of wearying you, to present some suggestions which I hope will be pertinent to the occasion, asking you to bear in mind constantly that the majority of the committee are the parties really responsible for the infliction.

Anything like a genealogical account of the family, desirable as it may be, will hardly be expected from me upon the present occasion. When I see before me so many men, better informed and more capable than myself to perform that duty, it would be an assumption in me to make the attempt. I once attempted to investigate the matter with reference simply to the descendants of Josiah. The result was somewhat amusing, yet it probably corresponds with that of many others, who have endeavored to trace out the Puritan families of New England. When one finds great uncles who cannot tell the names of their grandfathers; finds others who have ignorantly married their own relatives more or less remote, and begins to ask himself the questions, Whence? and Whither?—when one learns that the blood which flows in his veins has its counterpart in the veins of a whole section of the country,—the study of the various grades of relationship, and the tracing out of kith and kin is an employment which runs him into a scale of mathematical minutiæ which it is almost frightful to contemplate. The genealogy of the children of Israel might be traced with almost as much ease as that of some of the families of New England, because while you may know a Jew by marks which are not to be mistaken, your own next-door neighbor may be some tenth or sixteenth cousin, the fruit of some stock which has grown out of the knowledge of the original, while the

fibres which connected it with the parent root are so deeply
buried as to be almost entirely hid from the most careful
investigation.

In one respect we are fortunate. We know the founder
of our race in this country. His name has come down
to us as a fixed and certain landmark in our history. In
him we all meet. Starting at that point the years vanish,
—the various generations are forgotten,—names, distinctions and families for the moment disappear, and we meet
here to-day, a family of brothers and sisters, all at the old
homestead,—the children of Deacon Samuel Chapin.

The home of our honored ancestor was in this place.
Here he lived and died. There are records of various
lots of land granted to Dea. Samuel Chapin. I will refer
to but one entry:—"February 1, 1658.—Samuel Chapin
hath a house lot granted him from ye plantations, containing four acres more or less, Breadth 8 rods, Length
80 rods, abutting against the street East, and the Great
River West, bounded by Mr. Moxon, North ; by Thomas
Rives, South. Also on the same line before his house lot,
Eastward, all the wet meadow containing to the value of
about two acres more or less, and at the end of the
wet meadow lies a wood lot of four acres more or less,
Breadth 8 rods, Length 80 rods, running on the same
line bounded as the home lot is. Also over the Great
River a lot of four acres more or less abutting against the
Great River East ; and thence running in length Westward 74 rods. The breadth 8 rods. Bounded North by
Mr. Moxon, South by Mr. Rives." * * The location of
this home lot can be substantially identified. The earliest
account of the birth of any child of Samuel Chapin is
that of Japhet, in the year 1642. What was the father's
age at that time, or what was the date or place of his

3

birth, history does not inform us. All we know is, that
the earliest record of the birth of a child in this country
is as previously stated, and that he married Cicely. Who
was Cicely? This, of course, was her Christian name,
for her surname is covered by the dark mantle of more
than two hundred years, and the keen eye of no historian
of the family has ever been able to penetrate the mystery,
or help to answer the question. From well authenticated
records, we are enabled to say that in the order of seating
persons in the meeting-house, she sat in one of the most
honorable seats, along with Mrs. Glover and Mrs. Holyoke, who was a sister of William Pynchon, and that she
"was sick and dyed, Feb. 8, 1682," having survived her
husband six years, two months and twenty-eight days.
From these facts, and from the qualities which have been
transmitted in the heads and hearts of her bright-eyed
daughters of to-day, I am fully prepared to assert and
maintain that she was a practical, pious woman, who
proved a worthy partner to her husband, and who helped
to make the wilderness around them blossom like the rose.
The worthy deacon undoubtedly often pondered upon the
description of one like her in the last chapter of Proverbs;
and given as he must have been to making quotations
from Scripture, Cicely must have heard from his lips, just
often enough to make it acceptable, "She looketh well to
the ways of her household, and eateth not the bread of
idleness. Her children arise up and call her blessed; her
husband *also*, and he praiseth her. Many daughters have
done virtuously, but thou excellest them all." We pay
our tribute to this New England mother, and turn our attention to the first cis-atlantic male member of the family.

What do we know of him? What can we tell of the
venerable Deacon of the Valley of the Connecticut? We

19

arrive at our conclusions in this world from observation by the senses; from faith, which is the substance of things hoped for, and the evidence of things not seen; or from belief founded upon facts, from which we draw our own inferences. Of him personally, we of course know nothing. In him our faith is large and hopeful, while our belief of his worth and ability is gathered from the scattered relics which bear upon his history, and from the men and women whom I see before me.

In Allen's Biographical Dictionary the following statement is found: "Seth Chapin—Deacon—an officer in the Revolutionary war, died in Mendon, November 15th, 1833, aged 79 years. His grandfather, Joshua, came from Lancashire with a brother Gershom, who settled at Springfield. From these have sprung many ministers." In an obituary notice of Dea. Seth Chapin, published in the *Boston Recorder* nearly thirty years since, the following statement is found: "Dea. Chapin was of the third generation in descent from Joshua Chapin, Esq., one of the first settlers of Mendon, who contracted with the Indians for the town, the oldest town in Worcester County excepting Lancaster, and wrote the deed which is still extant. That ancestor, with two of his brothers, Gershom and John, emigrated from Lancashire, England, and it is believed no others of the name have ever settled in this country. Of these, John, the younger brother, not being pleased with his prospects, soon re-embarked to return, and was lost at sea. Gershom, not satisfied with the good soil of Mendon, sought a better country;" (meaning Springfield or Agawam.) The statements have puzzled more than one person who has attempted to investigate the family history. I deem it to be my duty to say distinctly and unequivocally that these statements are erroneous. John Chapin may have been lost at sea, but no Chapin with the Christian name of Joshua or Gershom ever came from Lancashire to Springfield, and no Chapin with the name of Joshua was one of the early settlers of Mendon. Dea. Seth Chapin referred to was the son of Ebenezer Chapin, who was the son of Seth Chapin, who was the son of Josiah Chapin, who was the son of Samuel Chapin, who was the identical Samuel Chapin who took the freeman's oath at Boston, in the year 1641; arrived in Springfield in 1642, and who is referred to in the life of William Pynchon, very properly called the founder of Springfield. William Pynchon, owing to some disaffection growing out of the publication by him of certain views upon theology, in a book which was denounced, condemned and burned, left Springfield and returned to England in the year 1652. Says the writer: "The General Court immediately appointed his son, with Elizur Holyoke, his brother-in-law, and Samuel Chapin, (the ancestor of all of that name in New England,) Commissioners to exercise the powers of Magistrates in Springfield." It may therefore be laid down as a fact well established and abundantly proved, and which all the evidence from records and traditions tends to confirm, that the Chapins in this country did not descend from three or four stout brothers, who, as the common saying is, "came over together," but that all of the name, unless they have adopted it either with or without legal authority, are the direct lineal descendants of Samuel and Cicely Chapin. The evidence is quite satisfactory, however, that when Samuel Chapin came to this country he brought with him a family of at least five children, and resided for a time in Roxbury. The records of Roxbury refer to him as having a family of eight persons; and the fact that

Josiah Chapin settled in the easterly part of the State, while the rest of the family took up their residence in Western Massachusetts, may be explained in part by the fact of that early residence in Roxbury, and by other facts to which I may hereafter refer.

So much being established, let us endeavor to ascertain as best we may some of the leading facts which will be likely to interest us as a family. In this connexion I may be allowed to say a word upon the subject of the family name. It is uncertain how long the name has been written in the form and with the termination which we now find. It seems to have been spelled at least four different ways:—Chapaen—Chapun—Chapen, and Chapin. The last has been the uniform spelling of the name in this country since the first half of the seventeenth century, and in this country it has been the uniform spelling of the name when written by any member of the family. Any one who has observed the habits of the people of New England two hundred years ago, in respect to the spelling of their names, will not be surprised at the slight variation which has taken place in our own. Uniform spelling, even of their own names, was not one of the idols which they worshipped, and if they, away from their early firesides, could resist the temptations of the devil, keep the foes of Puritanism away, and soul and body together in a region where cultivation had been a stranger, they toiled on with hearts resting upon the promises, and "builded better than they knew," while the exact orthography of the present day might have been at a discount; or they may have preferred to change the form of their own particular cognomen, as a more effectual means of breaking the last link which could bind them to that country where they had suffered persecution for opinion's sake, and where, in the eloquent language of James Otis, they had "plunged into the wave with the great charter of freedom in their teeth, because the faggot and the torch were behind" them.

The first that we learn of Samuel Chapin is that between the years 1636 and 1640 he with his family were residing in Roxbury, in the Colony of Massachusetts Bay. In this connexion it may not be improper to say, that in the History of Roxbury it is stated, that "It has been remarked that no people can boast of more honorable descent than those of Massachusetts," and it is also recorded that "the Roxbury people were of the best that came." The exact year of his arrival in this country cannot be fixed. It was probably in 1635 or 1636. The number of the family, as has been stated, is given as eight. If the list of the family which comes to us is the correct one, all of the children except Hannah (who according to the record was "born 10 mo. 2 day, 1644,—about 10 o'clock at night—and baptized the 8th day,") must have been born prior to the removal of the family to Springfield. The statement of the quantity of land owned by Samuel Chapin in Roxbury is too illegible to be deciphered. The records of Roxbury contain nothing relating to him of special interest. He probably remained there but a few years. In 1641, as already suggested, he took the freeman's oath in Boston, and thus became a citizen of the Colony, and entitled to vote. It sounds strangely to us, at this day, to hear that the essential qualification for being admitted as a freeman, or in other words, to the rights of a citizen, was to be a member of the visible Church. The Puritans were determined that, so far as they were concerned, the affairs of the Colony should be managed by men of their views of religion, and disregarding what seems to us the inherent right of private judgment; blind to what we consider the evil of combining Church and State; observing not the tendency to raise up a nation of hypocrites, they applied their principles with a naked rigor and accuracy which startle the men of this liberal generation. In the beautiful language of Mrs. Hemans:

"They left unstained what there they found;
Freedom to worship God;"

but unless a man chose to enroll himself as a member of the Church, he could not enjoy the privileges of a free citizen. At first, when nearly all persons in the colony were members of the church, the severity of the system was but slightly felt; but the pressure of these principles and practices in the end, undoubtedly produced much restiveness and dissatisfaction. The half way covenant, as it has often been called, was probably the legitimate result of this state of things. In a few words "It was held," by the synod called at Boston, "that all baptized persons were to be considered members of the church, and, if not openly dissolute, admitted to its privileges, except partaking of the Lord's Supper." One great object of this undoubtedly was to escape from the rigor which required a freeman of the colony to be a church member. The result of this half way covenant was considered so pernicious that in the end it was laid aside in all the Orthodox Congregational Churches. It is well stated by a writer upon those early times, "The history of the church is the history of the plantations. Without intending it, and indeed with principles in their full development essentially hostile to any connexion between the State and the Church, the Pilgrims so blended together religious and political institutions, that both religious and political liberty grew sickly and feeble from the unnatural union."

Without dwelling upon questions which are not likely to arise among us, we learn from the fact that our ancestor was admitted a freeman in 1641, that he was at that time a member of the church, and that prior to that time he could have taken no active part in the affairs of the colony. Had we, however, no other light to guide us, we might not be perfectly sure that he was a converted man; because human nature might not have withstood successfully the exclusion which resulted from the stern qualifications which were required by the Puritans, and much as we honor the fathers of New England, deeply as we reverence their true sincerity of purpose, sincerely as we believe, in the language of William Stoughton, that "God sifted a whole nation, that he might send choice grain into the wilderness," looking back, with silent admiration at their desperate conflicts with temporal and spiritual temptations, and their rigid banishment of many of the lighter graces of life, because to them their tendency was evil, we cannot help feeling that to some of their descendants their requirements might have been irksome, and their honest purity been tainted with what to some of us might have seemed to be intolerance.

But our theme is not the Puritans, but a Puritan. Arriving in Springfield in the year 1642, Samuel Chapin soon appears to be one of the most prominent and reliable citizens of the place, and soon after the organization of the church he was undoubtedly selected one of the first deacons thereof. From that time he is designated as Deacon Chapin.

This office of deacon has different meanings attached to it, in the different connections in which the office is held. First established by the Apostles, the office was held by men who distributed alms, provisions at the love feasts, and bread and wine to the communicants. In the Catholic church, the deacon is an ecclesiastic, the second in the sacred orders, and not allowed to marry. For some good reason, probably, no descendant of Dea. Samuel Chapin has ever been a deacon in the Catholic church. In the English church, deacons are ecclesiastics with limited jurisdiction, and in the Presbyterian and Independent churches their particular duty is to distribute the bread and wine to communicants. This office, however, has usually been conferred upon men, whose gift of speech has enabled them to assist in the devotional exercises of the church, and whose religious character has marked them as suitable examples for those who are attempting to lead a Christian life.

Our ancestor evidently did not confine himself to the strictly technical duties of a Puritan deacon. In November, 1655, a vote was passed "to allow Deacon Wright, Deacon Chapin, Mr. Holyoke and Henry Burt £12, for their past services in the Lord's work on the Sabbath, to be distributed by the selectmen, and that in future they would allow at the rate of £50 a year till such time as they should have a settled minister, to be distributed and ordered by the selectmen."

In 1656, "Mr. Holyoke and Henry Burt were appointed to carry on the work of the Sabbath; and if they were disabled, Dea. Chapin was to supply their place."

"In November, 1657, Mr. Holyoke was chosen to carry on the work of the Sabbath once every Sabbath day, which he accepts of. Mr. Pynchon is made choice of for one part of the day once a fortnight, to which he will endeavor to attend, sometimes by reading notes and some-

4

times by his own meditations, till March next." "Dea. Chapin and Henry Burt were selected to carry on the other part of the day once a fortnight, to be allowed at the rate of £40 a year."

We also learn from the records that, on "February 23, 1662, the order of seating persons in the meeting, Goodwife Chapin " (Cicely, of course,) " is to sit in ye seat along with Mrs. Glover and Mrs. Holyoke."

From the foregoing statements, we may reasonably infer that both husband and wife possessed the confidence and respect of the religious community, and the additional facts which are found upon record, authorize the additional inference that Dea. Samuel Chapin acted a very important part in the civil and municipal affairs of Springfield.

In 1662, January 26, Samuel Chapin and five other persons were authorized by "general vote and consent of ye plantation, to lay out the lands both of upland and meadow, on ye other side of ye Great River, where ye Indians live, and all the meadow on Agawam so far as shall amount to one hundred and fifty acres."

In the records of the year 1644, we find the following : " Springfield, the 26th of the 7th month, 1644.—It is this day agreed by general vote of the town, that Henry Smith, Thomas Cooper, Samuel Chapin, Richard Sikes, and Henry Burt shall have power to direct in all the prudential affairs of the town, to prevent any thing they shall judge to be to the damage of the town, and to order any thing they shall judge to be for the good of the town ; and they or any three of the five shall have full power for a year's space ; and what they or any three of them shall order shall be of full power and virtue. Also, to *hear complaints*, to *arbitrate controversies*, to lay out highways, to make bridges, repair highways, especially to order the making of the way over the musky meadow, to see to the securing of the ditches, and to the killing of wolves, and to the training up of the children in their good ruling, or any other thing, they shall judge to be to the profit of the town."

While William Pynchon remained in Springfield, he acted as magistrate. For a short period after his departure for England, in the year 1652, the duties of magistrate were performed by his son-in-law, Henry Smith. After his departure for England, in 1653, the General Court, as I have already stated, appointed John Pynchon, Elizur Holyoke and Samuel Chapin, Commissioners to exercise the powers of Magistracy in Springfield. Their duties were extensive, delicate and important. They were in fact the judicial officers of the place. They exercised civil, criminal and probate jurisdiction. They heard and determined suits brought for the recovery of debts, and of damages for injuries inflicted. They administered the penalties for card playing, and for a numerous catalogue of other offences. They assigned her just share to the widow, and guarded the interests of the orphan. The ancient volume in which their decisions are entered, is one of the precious relics of those early times. To us there is one entry of more than ordinary interest. It is the name of Samuel Chapin, entered in his own hand writing. It is a remarkable specimen of ancient chirography. No man could have written that name who was not a man of character and education. It justifies the remark of one writer, that "Deacon Samuel Chapin was a man of distinction," and that of another writer, that "he was well qualified for public business." That the name referred to was written by him, is demonstrated by the fact that the document requires his own personal signature, and by his own hand writing, as it appears in other documents purporting to have been written by himself. And here I may be allowed to say that, strange as the idea may seem, there are those who sincerely believe that a man's handwriting, when he is in full health, is a true index of his character. No man who was a sneak or a hypocrite, ever wrote a clear, round, legible hand, while many a shriveled, insincere soul has been unconsciously though indelibly stamped upon the written page, by the hand which has unwittingly furnished evidence against it. No descendant of Samuel Chapin who has seen his handwriting, need ever to be ashamed of his progenitor.

It may not be uninteresting to state, briefly, some of the other official relations which our common ancestor sustained towards the people of this place. May 19, 1645, Samuel Chapin and six others, "with the joint assent of all the plantations at a public meeting after sufficient warning, were appointed to divide the town into equal parts for estates and persons." November 2, 1647, Samuel Chapin and four other persons "were chosen for ordering the prudential affairs of the town." In other words, they were selectmen. November 6, 1648, the same persons were chosen selectmen or townsmen. November, 1649, the same persons are supposed to have been chosen townsmen. November 5, 1650, the same persons were chosen townsmen. November 4, 1651, Samuel Chapin was one of the townsmen. September 14, 1652, Samuel Chapin was one of the committee to purchase Mr. Moxon's house to remain for ye use of the ministry to posterity. (Mr. Moxon, who had been the minister, had concluded for some cause to return to England.) November 2, 1652, the same persons were chosen townsmen. February 5, 1660, Samuel Chapin was chosen one of the selectmen. "August 14, 1662, Dea. Samuel Chapin was chosen for ye commissioner to join with ye selectmen in making ye county rate and to do therein as ye law enjoins."

October 12, 1667, we find a record of a conveyance from Samuel Chapin to his son Japhet Chapin, of his housing and certain parcels of land described in the conveyance, and November 11, 1675, Samuel Chapin, deacon of the church at Springfield, died.

If I am not mistaken as to the feelings of this audience, I think that we are all prepared to assert that our honored ancestor in his day and generation was every inch a man. He was no miser who starved himself for the sake of his pocket. He never skulked around from place to place about the first day of May, in order to escape from contributing his honest proportion of the expenses of supporting the institutions of society. He was not a rich man, who contented himself with giving merely the crumbs to some poor Lazarus at his gate, while the dogs should have put him to shame by their acts of kindness and good will. He was no family tyrant, in whose presence his wife held her breath and his children dared not say that their souls were their own. He was not one of those close fisted husbands, who cannot imagine what a woman *can* possibly want of a five dollar bill. He was not one of those contracted creatures who will neither borrow nor lend, but live like a tortoise with his head in his shell. He did not pull in his own latch string, and selfishly enjoy the good things of this world, without bidding his neighbors and friends participate. He was no Pharisee, shouting to each sinner about him, " Stand aside, for I am holier than thou." He never violated his conscience for the sake of servile obedience to a wicked law. He never turned his servants from his door without promptly paying them their honest earnings. He winked not at tyranny, meanness or injustice. The poor and feeble never appealed to him in vain, and the world was not made worse by his having lived in it.

I have stated what he was not. I am as clear and decided in my own mind as to his positive qualities, as I am to his negative ones.

We look back through the centuries, and we seem to discern the figure of a firm, square built man about six feet in height, who had that consistent element of character which led him always to say what he meant, and to mean what he said. He was one of the positive, earnest and efficient men of those early times. In whatever department of duty his course was to be chosen, he was sure to succeed. Had he been a farmer, the best of stock and fruit and grain and vegetable would have flourished and fructified upon his broad acres. Had he turned his attention to theology he would of course have made a doctor of divinity. Had he made the law his profession, he would have been an ornament to the bench or the bar. Had railroads been in existence, he would very probably have been a Superintendent of the Cheshire or a President of the Western Railroad. Had he been a landlord, his hotel would have been as popular as the far-famed Massasoit House. Had Springfield been a city, he would undoubtedly have been its chief magistrate ; and if he happened to be a member of the humble church in the village where he had found a home, willing or not he was bound to be a deacon thereof. (Note A.) Venerable patriarch! we need no word of thine to assure us of thy character. Thy descendants scattered over the length and breadth of this fair land have been living witnesses, coming down through the centuries, to exhibit to those around them the traits which they have inherited from thee. The virtues of the father may live in the children. In our case, our follies and sins must certainly be our own. We gather in grateful reverence around thy honored dust, and bring our garlands of love and good will, to shed fragrance upon thy memory. If thy disembodied spirit takes cognizance of terrestrial events, surely this goodly gathering of thy family must fill thee with honest pride, while we thy children shall carry to our homes the fancied lineaments of thy face and form. (Note B.)

What changes have been wrought here, since the day when Samuel Chapin settled in the valley of the Connecticut! This silent river indeed flowed onward toward the sea,—these beautiful hills reared their summits toward the sky,—these rich intervales, pregnan twith the wealth of coming generations, spread their broad surface to the sun,—the natural form and structure of this territory were the same as at this hour, but all else how changed! This beautiful valley, whose fame is in the mouths of all who have ever looked upon it, was a wilderness. The hand of art had never changed this flowing river into a means of industry and skill. The roar of its artificial waterfalls was not heard in the distance. The coffers of the general government had never been drawn upon for its civilization and adornment. The steaming steeds shot not like fiery meteors across the country. The lightnings were not then the winged messengers for man to hold converse with man. No well-built city here flourished and grew rich, but the quiet settlers labored in rural simplicity, unconscious that on the spot where they were toiling so faithfully and well, a city would arise to rival in beauty, intelligence and prosperity many a proud municipality of the old world. How little do we realize the trials and dangers of the men and women of that early settlement! To-day a call comes up from the capital of this great country, and almost before the wires have delivered their message, New England troops are on their march to protect the honor of our flag. In 1642, whole neighborhoods might have been swept away by war or pestilence, and weeks and months have elapsed, without aid, succor, or even a communication of the fact.

The early settlement of Springfield, or Agawam as it was then called, in 1636, one hundred miles distant from the first settlement in Massachusetts Bay, has at times been a cause of wonder to those who have reflected upon the subject. The Plymouth colony was first settled only as early as 1620; Salem, in 1628; Lynn, in 1629; Boston, in 1630; Concord, in 1635; and Worcester, in 1673. Why, then, should the first settlers in this valley have wandered so far away from their friends, and left such a space of wilderness between them? We must bear in mind that the fathers of New England were determined to have room enough, and that they were very much in the habit of doing things in their own way. It may be that they saw in the great river an easy avenue to the ocean, and felt that God had given to them a highway which they would do well to occupy. The Indians were comparatively well disposed. They had not learned to hate the white man, or to imitate his vices. The men who had studied in the school of oppression in the old world, had learned what it was to escape from persecution, and, wearied by their lessons, were determined to enjoy one breath of freedom, although the wild beasts were their neighbors, and the wind, coursing through the trees of the wilderness, sang their lullaby or their requiem.

In examining the early history of the country, one is struck with the importance which was attached to meadow land. The present value of much that is termed meadow is very different from the value of the meadows of that early day. The natural grass, furnishing food for beast without the necessity of cultivation, made such property very desirable, and no farm was considered well balanced which had not a portion of such land. This is not the occasion to discuss the qualities of different kinds of meadow land, or to explain the difference between the magnificent intervales of the Connecticut valley and the meadows which abound in some other parts of the State; but to me it is perfectly evident that these fertile meadows or intervales which year by year teem with such splendid grass or luxuriant green, were among the causes which operated to bring about the settlement of Springfield at so early a period.

Not being a native or a resident of this beautiful valley, the speaker has a right to say that the first settlers here were not only brave, God-fearing men, but that some of them possessed that intuitive gift of foresight which enabled them to look forward, through the centuries, upon what we now behold. It is true that they contemplated nominally but forty families. They saved their lots running from the street down to the river. They saved the woodlands lying on the east side of the street and extending up the slope now radiant with beauty. They had an eye to the intervale lying opposite each lot, and on the west side of the great river, as soon as they could provide means for reaching it. They indeed said nothing about this being the location of a great inland city, but it is difficult to believe that visions of the wealth and position of those who should come after them did not flit through the

5

brains of some of those men so outwardly practical. They undoubtedly saw with the eye of faith and hope, and sowed the seeds of a golden harvest, which their sons are now reaping and gathering in. These settlers here were the pioneers of their day—the traveling progressives who endured far more than we can now realize, to lay the solid foundations of a community where the virtues and graces of life should find a welcome and a home.

Thus it is ever with the true pioneers. The advocate of a true idea may be separated from his friends by an almost impenetrable wilderness of confusion and darkness which for a time prevents a proper appreciation of him or his motives. The wild beasts of misrepresentation may surround him. The ground, all untrod and uncultivated, almost sinks beneath his feet, and he may seem to be hopelessly separated from those most near and dear to him. But to him all is bright and hopeful. An unseen light illumines his pathway. Visions of future glory and success give peace and comfort to his dreams, and many a man, even in his own day and generation, realizes a fruition of principles once cast out as evil, yet fondly cherished by himself, and enjoys the interior consciousness and the outward evidence of having been devoted to the truth and the right. But whether the fruit of correct opinions is to be gathered or not in his own day, many a true man has gone down to his grave as absolutely certain of the result, as any early settler in the valley of the Connecticut could have foreseen the present glory of this garden of the old Bay State.

The children of Dea. Samuel Chapin were Catharine, Sarah, David, Henry, Josiah, Japhet, and Hannah. The first five were probably born in England or Wales, Japhet in Roxbury, and Hannah in Springfield. It appears that

Jane Chapin, a daughter of Shem and Deborah, was born in Boston, September 16, 1665. As nothing more is known of the family, it is presumed that either the entry is incorrect, or that there are no descendants of Shem and Deborah in this country. Whether he was a son, brother, or sustained any other relation to Dea. Samuel Chapin, does not appear. The same remarks apply to John Chapin, who is referred to in the records of the Colony of Massachusetts Bay, in New England, at the Court of Assistants holden at Newtown, July 1, 1634, in connexion with a movable fort, and a meadow which John Chapin hath already mowed. Much effort has been made to trace the family in this country, and it has uniformly resulted in the conviction, that every person who has inherited the name is a direct descendant of Dea. Samuel Chapin.

Catharine, the oldest child of Samuel and Cicely Chapin, married Nathaniel Bliss, ninth month, twentieth day, 1646,—Thomas Gilbert, January 30, 1655,—and Samuel Marshfield, December 8, 1664, and became the mother of eleven children—to wit: four of the first family, four of the second, and three of the third family. If I have not forgotten, it was recently stated in the New York *Independent* that Catharine, a daughter of Dea. Samuel Chapin, married Nathaniel Foote. With all due respect for a sheet usually so accurate and true, it becomes my duty to correct this statement. If Catharine Chapin had not already had three husbands, the deacon would have undoubtedly objected to her marrying Nathaniel Foote, if for no other reason, on account of the difference in their ages, but I doubt not that he would have cheerfully assented to the marriage of his granddaughter Margaret Bliss to Nathaniel Foote, especially if he could have foreseen that among the results of that union would be Senator Foote, of the State of Vermont, and Rev. Henry Ward Beecher, the preacher of the Plymouth Church at Brooklyn, New York.

Sarah Chapin married Roland Thomas, second month, fourteenth day, 1647, and became the mother of thirteen children. Hannah Chapin married John Hitchcock, September 27, 1666, and became the mother of nine children.

The tradition is that they all married very respectably, while the facts which are known in reference to the families which have descended from them, not only do great credit to the discernment of the daughters in their selection of husbands, but justify the assertion that they inherited in large measure, the virtues and accomplishments of both their father and mother.

Although the statement made by one who was collecting the genealogy of his family is true, when he says, "I argued that a man's daughters are as much related to him as his sons," yet it is natural, upon an occasion like this, that more attention should be devoted to the male members of the family, because this is their only chance, while the females, who have entered into other connexions, may find their own histories written, and themselves living over their lives again in the memorials of some other family tree, upon which they have become engrafted by marriage.

Any one who has examined the facts which have been collected by our friend from Willimansett, and has seen the space which they fill, will begin to understand what must be the size of a volume which should embrace within its pages the names of those who, if living, would be included in the terms of the call for this meeting. Beginning with the year 1646, when Catharine Chapin married Nathaniel Bliss, and counting as our family all who come within the meaning of the phrase "descendants of Dea. Samuel Chapin," we find the multiplication table fading into insignificance, and the head growing giddy, while we attempt to calculate the number and extent of the progeny. Take a single example. We meet somewhere in our investigation the statement that Dea. William Rockwell, supposed to be the ancestor of all the Rockwells in this country, married Sarah Chapin. Although she was not the Sarah Chapin who married Roland Thomas, yet if a later Sarah married a man who was the common ancestor of a family so numerous and respectable as that of Rockwell, where shall we limit the results of the marriages of Catharine with Nathaniel Bliss, Thomas Gilbert and Samuel Marshfield; Sarah with Roland Thomas, and Hannah with John Hitchcock, when the united children of the three daughters, nearly two hundred years ago, numbered thirty-three?

David Chapin was the oldest son of Samuel, about whom we know any thing. He was admitted a freeman at Springfield in the year 1649, and on the twenty-ninth day of the sixth month, 1654, was married to Lydia Crump of Boston. According to the records, two children, Lydia and Caleb, were born to them in Springfield, and five, Sarah, Hannah, Ebenezer, Jonathan, and Union, were born in Boston, where David Chapin, the father, died in 1672. Less is known of him and his family, than of any other of the children of Samuel Chapin. November 4, 1656, he was chosen sealer of weights and measures in Springfield, and took the oath. Soon after that time, he removed from Springfield, and, in the language of one writer, "no more heard of him." This statement is erroneous. He removed to Boston, where he had five children born, and died there. We have not been able to follow the fortunes of his family. He died comparatively young, and his family is scattered or extinct. We know enough of him, however, to state positively that he had the respect of his townsmen and associates, and although he died a hundred miles away from the home of his early manhood, at a time when the dark wilderness covered most of the intervening territory, we doubt not that he died the death of a good man, and that the tears of a family and friends who loved him paid their unquestioning tribute to his memory.

The remaining children of Samuel and Cicely Chapin were Henry, Japhet and Josiah. Between Henry and Japhet, we have reason to suppose that there was remarkable sympathy and affection, because although Samuel Chapin in the language of an instrument executed by him October 12, 1667, " declares that he had upon his son Japhet's marriage, given and granted unto him, his said son, full right, title and interest unto the one half of all his, the said Samuel Chapin's housing" * * " certain parcels of land hereafter mentioned, and *all* his, the said Samuel Chapin's housing in Springfield which he possesseth, and *all* the said parcels of land hereafter mentioned, after the decease of him, the said Samuel, and his wife," we find that these two brothers, after a few years, left the village of Springfield, removed into what is now known as Chicopee, and which was then a howling wilderness. In the language of another, " Here was the undisturbed lair of the wild beast, and the savage warrior found a safe retreat from his pursuer in the tangled thicket. Hither the two brothers Japhet and Henry came and planted themselves down in the midst of the forest. Japhet built his house at the north end of what is now Chicopee Street."

* * " Henry located somewhere towards the lower

end of the street. They together, at that time, owned most of the land lying between Chicopee River and Willimansett Brook, and extending some distance eastward on to the plain." Here they reared families of children, Japhet having been the father of ten and Henry of five children. Six of the children left by Japhet, and two of those left by Henry were sons. The eight sons of Japhet and Henry had eighty-seven children born to them, giving to Japhet sixty-five and to Henry twenty-two grandchildren. We thus find that the family name soon became very prominent and numerous among the inhabitants of Springfield, which comprised originally a territory nearly twenty-five miles square, including what is now Springfield, West Springfield, Westfield, Southwick, Suffield, Enfield, Longmeadow. Somers, Wilbraham and Ludlow. During a number of years after the removal of Henry and Japhet to Chicopee, those members of the Chapin family who retained the name in Western Massachusetts, were found principally in what is now Chicopee, although the name appears among those of the original inhabitants of Wilbraham and Ludlow. We find that of twenty-seven persons taxed as belonging to the parish in Chicopee in the year 1753, twenty bore the name of Chapin, and that in the church in Chicopee, consisting of forty-three persons, in the year 1752, thirty-two bore the same surname. Gradually they have become scattered through the various avenues of industry and success in this country, until it has become almost impossible to follow their wanderings, or to find their resting places. They are so widely diffused that it requires more time than the speaker has at his command, to trace the true line of separation between the families of the sons of Henry and Japhet and Josiah. It may be laid down as a general rule, (subject like all general rules to exceptions,) that if a person of the Chapin blood hails from Western Massachusetts, that person is a descendant either of Catharine, Sarah, Hannah, Japhet or Henry, and if a person of the same blood hails from eastern or central Massachusetts, that person is a descendant of Josiah. If one hails from Rochester in the State of New York, he is probably a descendant of Japhet, and if from Canandaigua in the same state, he is probably a descendant of Josiah. Great credit is due to Dea. Orange Chapin for his skill and industry in collecting and preserving so much of our history. Let us hope that a similar spirit may animate some other member of our wide spread family, and lead him to complete its genealogy.

The characters of Japhet and Henry seem to have had many points of resemblance. The writer from whom I have already quoted, says of them, "These men had been faithfully trained up from their childhood in the ways of virtue and religion by their pious father, and evinced in their lives that these parental labors were not in vain." They were undoubtedly the leading men in their section of the town, and in the year 1689, Henry Chapin was a representative from the town of Springfield in the General Court at Boston. Tradition says of him, that he was in the early part of his life impressed on board a British man-of-war,—that he was in a severe engagement with the Dutch,—that he afterwards commanded a merchant ship and made several voyages between London and Boston, and that after residing for a time in Boston, he finally settled in Springfield. The record shows that "March 26, 1660, Townsmen Meeting—Henry Chapin is admitted an inhabitant of this town, and Dea. Chapin acknowledgeth himself bound to the town treasurer in a bond of £20, to save the town harmless from any charge for the said Henry Chapin." He died August 15, 1718, having resided fifty-eight years in the town of Springfield, and was probably somewhere between eighty and ninety years old at the time of his death. Japhet, for a few years after his marriage, resided in Milford, in the State of Connecticut, then removed to Springfield, and died in Chicopee village, February 20, 1712, at the age of seventy years. He was the youngest son, and seems to have been a kind of Benjamin in the family of his father. To him the old man deeded the homestead, retaining the use of one half thereof during his natural life, and no evidence or tradition of dissatisfaction on the part of either brother or sister, has come down to us to mar the beauty of that affectionate family circle, while Japhet himself, in spite of the dangers which oftentimes surround the youngest son of a devoted and indulgent parent, withstood the temptations which beset him, and became "particularly distinguished for his devoted piety."

The mantle of the first deacon seems to have fallen in large measure upon his descendants. Benjamin Chapin, a son of Henry, and David Chapin, a son of Japhet, in the year 1752, each at the age of seventy years, became deacons of the church in Chicopee. Benjamin died in 1756, aged seventy-four years, and David died in 1772, aged ninety years. Edward Chapin, a descendant of Japhet, became deacon of the same church in 1773, and died in the year 1800, aged seventy-six years. Giles S. Chapin, a descendant of Henry, became deacon of the same church in 1837, and Orange Chapin, a descendant of Japhet, became deacon of the same church in 1840. Aaron Chapin, a descendant of Japhet, became deacon of the First Congregational Church in Hartford, Ct., and

6

died, universally beloved and respected, December 25, 1838, in his eighty-sixth year. Japhet Chapin, a descendant of Japhet, became deacon of the Baptist Church in Buckland, Mass., and died April 22, 1833, aged 76 years. Alpha Chapin, a descendant of Japhet, is deacon of the Presbyterian Church in Ogden, N. Y. Curtis Chapin, a descendant of Japhet, is deacon of the Orthodox Congregational Church in Bernardston, Mass. Caleb Chapin, a descendant of Japhet, is deacon of the Unitarian Congregational Church in Bernardston, Mass. Enoch Chapin, a descendant of Japhet, is deacon of a church at South Hadley Falls. Samuel W. Chapin, a descendant of Japhet, was deacon of the Orthodox Congregational Church in Bernardston, Mass., and died November 4, 1851, aged sixty-four years. David Belden Chapin, a descendant of Japhet, is deacon of the Congregational Church in Newport, N. H. Louis Chapin, a descendant of Japhet, is an elder in the Second Presbyterian Church in Rochester, N. Y. Edward Colton Chapin, a descendant of Japhet, is a deacon or elder of the Thirteenth Street Presbyterian Church, New York. Moses Chapin, a descendant of Japhet, is an elder in the First Presbyterian Church in Rochester, N. Y. Henry Chapin, a descendant of Japhet, is deacon of the Methodist Pynchon Street Church in Springfield. Japhet Chapin, a descendant of Japhet, is deacon of the Congregational Church in Antwerp, N. Y.

So much for the deacons. Let us turn for a moment to the clergymen:—

Rev. Calvin Chapin, D. D., a descendant of Japhet, was president of Union College, and died at Rocky Hill, Wethersfield, Ct., March 17, 1851, aged eighty-seven years. "A most faithful and excellent minister; a good scholar, and a wise man." Rev. Pelatiah Chapin, a descendant of Japhet, formerly of Chesterfield, died in New Hampshire about the year 1836, aged ninety years. Rev. Reuben Chapin, a descendant of Japhet, born in Somers, Ct., September 5, 1778, prepared for a Congregational minister, but was incapacitated for preaching by a serious accident. Rev. Seth Smith Chapin, a descendant of Japhet, is an Episcopal clergyman, in Marshall, Mich. Rev. Aaron Lucius Chapin, a descendant of Japhet, graduated at Yale College in the year 1837, and is now president of Beloit College. Rev. Nathan Colton Chapin, a descendant of Japhet, has preached to the First Congregational Church in LaCrosse, Wis. Rev. Daniel E. Chapin, a descendant of Henry, a member of the New England Methodist Conference, has been stationed at various places, represented the town of Webster in the Massachusetts Constitutional Convention of 1853, and the city of Worcester in the Legislature of 1855. Rev. Edwin H. Chapin, a descendant of Japhet, is the eloquent and devoted preacher at the Fourth Universalist Church, Broadway, New York. Rev. William A. Chapin, a descendant of Japhet, born at Newport, N. H., 1790, was settled in Greensboro, Vt., and died November 27, 1850, aged sixty years. Rev. Jason Chapin, a descendant of Japhet, born at Newport, N. H., 1801, was settled in Geneseo, Henry County, Ill., and died September 11, 1846, aged forty-five years. Rev. Henry B. Chapin, a descendant of Japhet, settled in Trenton, N. J. Rev. Augustus L. Chapin, a descendant of Japhet, born at West Springfield, January 16, 1795, was settled at Lexington, N. Y. Rev. Lyman D. Chapin, a descendant of Japhet, graduated at Amherst College in 1858, and has recently been ordained a missionary. Rev. Walter Chapin, a descendant of Henry, graduated at Middlebury College in 1803, became a tutor at the same College, preached at Woodstock, Vt., and died in 1827, aged forty-eight years. Rev. Alonzo B. Chapin, a descendant of Japhet, born March 10, 1808, left a fair practice in the legal profession for the study of theology, became editor and publisher of the *Chronicle of the Church;* rector of the Episcopal Church at West Haven; rector of St. Luke's Church, South Glastenbury, and editor of the *Calendar* at Hartford, where he died, July 9, 1858, fifty years of age. He was a diligent student, and a good writer. Rev. Dennis Chapin, a descendant of Japhet, graduated in 1837, and preaches in Canada. Rev. Percy Chapin, a descendant of Japhet, born April 29, 1783, preached nearly thirty years in Pownal, Me., where he died, aged fifty-six years. Rev. Horace B. Chapin, a descendant of Japhet, born December 3, 1791, preached at Amherst, Mass., and Lewiston Falls, Me., and died at the age of forty-nine years. Rev. Asahel Chapin, a descendant of Japhet, is now settled in Vinton, Ind. Rev. Henry M. Chapin, a descendant of Japhet, born at Woodstock, Vt., April 21, 1825, is now preaching at Markclean, Marquette County, Wisconsin. Rev. Solomon Chapin, a descendant of Henry, born June 2, 1831, graduated at the Wesleyan University, and is a member of the New England Methodist Conference.

It is an interesting fact, that of these twenty-one clergymen three only are found to be of the family of Henry, and eighteen of the family of Japhet; and of the eighteen deacons, two only are found to be of the family of Henry, and sixteen of the family of Japhet. Upon further investigation, I am inclined to hazard the opinion, that although the sacred offices of Minister and Deacon are found more abundant in the descendants of Japhet than in those of Henry, more than their proportionate share of this world's business, honors and emoluments, has found its way among the descendants of Henry. How much the fact that Japhet was particularly distinguished for his devoted piety, and the fact that Henry fought the Dutch and commanded a merchant ship which made voyages between London and Boston, have had to do with the religious or business traits of their descendants, it would hardly become a member of the legal profession to discuss upon an occasion like this. The question may be metaphysical, physiological, moral or intellectual. It is beyond my skill, and I leave it to those who are fond of explaining the idiosyncracies of mankind to elucidate it at their leisure.

It may not be improper, however, for a member of the stock of Josiah Chapin, to express his gratification at the qualities of mind and heart which have prevailed in the innumerable cousins of almost every grade of remoteness, who have belonged to the wide spread families of the other children of the venerable deacon. Their names are found in every decent department of human society. Farmers, merchants, mechanics, physicians, ministers, deacons, lawyers, magistrates, mayors of cities, selectmen of towns, judges, delegates to constitutional conventions, senators and representatives to Congress, senators and representatives to the General Court, presidents of colleges, railroads, banks and insurance companies ; school teachers, professors, missionaries, &c., &c., they commend themselves as almost universally respectable, with scarcely one perceptible stain to cast a single dark shadow upon the multitudinous picture.

I turn briefly to some facts relating to the branch of Josiah Chapin, the remaining son of Dea. Samuel Chapin. Josiah was born in the year 1634, married in 1658 aged twenty-four years, and died in 1726 aged ninety-two years.

The following statement is taken from an entry made by him in his own handwriting: "In the month of November, 1658, I was

I. Married to Mary King, in Weymouth, who died May 30, 1676.

II. I was married to my second wife, Lydia Brown, at Ipswich, ye 20th of September, A. Dom. 1676, who died October 11, 1711.

III. I was married to my third wife, Mehitable Metcalf, in Dedham, the 22d of June, A. Dom. 1713."

In the absence of the following entry upon the records of Springfield, I should say that Josiah Chapin uniformly resided in eastern or central Massachusetts, from the time of his marriage to the time of his death. "August 1, 1663, Josiah Chapin is by the selectmen admitted to be an inhabitant of this town." His father gave the usual bond of indemnity. The records show that Mary Chapin, a daughter of Josiah, was born in Braintree, August 27, 1662, and that Deborah Chapin, another daughter of Josiah, was born in Braintree, June 16, 1664. Therefore if he did come to Springfield to reside, it must have been for a very short period of time. The chances are that his wife did not think it best to come at all, and having a sensible wife with a will of her own, he like a sensible man chose to follow her judgment.

He became the father of fifteen children. Of the children of the first marriage, Samuel was born in Weymouth, and John, Mary, who married Joseph Adams, grandfather of John Adams, second President of the United States, Deborah, Josiah, Shem, Seth, Joseph, Henry, Ephraim and a second Deborah, were born in Braintree, the first in 1659, and the last in 1675. Of the children of the second marriage, Lydia, Sarah and David were born in Braintree, and Hannah was born in Mendon, the first in 1677, and the last in 1684.

From the foregoing statement we learn that Josiah Chapin, obeying the scriptural injunction, went forth and peopled the earth, and that if his descendants like his son Seth, who also became the father of fifteen children, had all followed his example, this family gathering would have been vastly more crowded than it is upon the present occasion.

Josiah Chapin removed from Braintree to Mendon, and purchased and occupied the estate which in later days has been known as the Simeon Doggett place, situated about half a mile north of the Town Hall in Mendon. He was one of the first settlers of Mendon, and during nearly half a century was the leading man of the place, and acted a very important part in the management of its affairs. This town originally included within its limits not only its present territory, but what is now Milford, Blackstone, Uxbridge, Northbridge and a part of Upton. One by one the territory included in each of these towns has been assigned to a distinct and separate organization, and from having been a large town, and of great relative importance, Mendon is left small in territory, and apparently deficient in the elements of growth and prosperity. Wishing to investigate the family history, a few weeks since, I naturally sought there for some one of the family name, and found to my surprise, that even the assessors of the town were unable to furnish one solitary specimen, within the present municipality. The early records, however, are full of interest.

It appears that Samuel Chapin, of Springfield, acquired the right to take up certain lands in Mendon, and that the settlement of Josiah in that town is probably in some way connected with the same.

"At a General Court held at Boston the 18th of May, 1664. In answer to the petition of Samuel Chapin of Springfield, humbly desiring the favor of this Court to grant some lands in reference to service done. The Court judgeth it meet to grant him two hundred acres of land where he can find it not formerly granted to town or person."

By an instrument bearing date June 24, 1668, Samuel Chapin, "not having hitherto pitched upon or taken up any land by virtue of said grant" * * * "makes over" to his son Josiah Chapin, all his right, title, interest or privilege in the same. At a General Court held at Boston, 19th of May, 1669, the land was laid out in Mendon, and June 11, 1717, by the House of Representatives was allowed and confirmed to Josiah Chapin, as the two hundred acres of land granted originally to Samuel Chapin.

Josiah Chapin settled in Mendon between the time of the birth of his son David, in Braintree, November 11, 1680, and the 18th of January, 1682; because on the latter day, at a General Town Meeting, the town chose Sergeant Josiah Chapin, James Lovett and Samuel Read a committee for to see that a frame house eighteen feet square, was built for Mr. Rawson (the minister).

July 21, 1682, Sergt. Josiah Chapin and others were chosen "for to treat with Mr. Rawson and to renew our hold of him for his continuance with us for the future, and do give them full power to act in the behalf of the town as if themselves were present, and to rest satisfied with what they do."

October 25, 1682, Sergt. Josiah Chapin was chosen a surveyor "to lay out the remainder of the swamp lots, or any other land not laid out."

It is interesting to observe with what tenacity, in those

early days, they clung to the practice of giving to one his title, civil, military or religious. They might frequently fail to attach to a man his Christian name; his title was rarely, if ever, omitted. Josiah was yet but a Sergeant, but it was a title of honor, and must not be neglected.

In 1682, Sergeant Josiah Chapin was Chairman of the Selectmen, and in the same year, according to the record, "for the coredgement of Sergeant Josiah Chapin to build a saw mill he is authorized to take up the land that is due to him on the other side of the brook, which is eighty acres." In 1683, 1684, 1685 and 1686, Sergeant Josiah Chapin was one of the Selectmen. In 1687 and 1688 Ensign Josiah Chapin was one of the Selectmen. In 1689 Lieut. Josiah Chapin was one of the Selectmen; and in the same year "by a full vote was chosen and fully authorized to consult, advise, join and give his assistance with the honorable Council for safety of the people, and conservation of the peace." The same year he was elected Representative to the General Court. In 1690 Lieut. Josiah Chapin was one of the committee for building the meeting house. In 1691, Lieut. Josiah Chapin was Chairman of the Selectmen. In 1692, Capt. Josiah Chapin was Chairman of the Selectmen. During the same year his tax for the support of preaching was larger than that of any other person in town, except one. In 1693 and 1696, his tax for the same purpose was the largest in town. From the strongest evidence, it is easy to arrive at the conclusion, that in spite of being engaged in public business during forty years of his life, his means increased in a satisfactory proportion, and that his interest in the land was very extensive.

In 1693, Capt. Josiah Chapin was chosen clerk of the market. Not having been chosen one of the Selectmen,

7

Capt. Chapin and three others were chosen " to be a committee to give instructions to the Selectmen for the year ensuing." The instructions are upon record, and are very sensible and judicious. During the same year it is recorded that " whereas Capt. Josiah Chapin hath served the town as a representative, and " (the town) " falleth ten pounds in his debt, it passed by vote he should have and enjoy four-score and ten acres of land." During the same year he was chosen " to assist and join with the town clerk to record the General Court's grant." In 1694, 1695, 1696, 1697, 1698 and 1699, Capt. Josiah Chapin was chairman of the Selectmen. In 1700, he was chosen " a committee to give allowance on next division of land." In 1701, 1702, 1703, 1704 and 1705 he was chairman of the Selectmen. In 1703, it was voted that " Capt. Josiah Chapin should have the remainder of the money that was due to the town from Jonathan Richardson for his lot, for part of what is due to him for his services at the General Court." In 1706, a new and peculiar trust was committed to him. In some controversy with the town of Holliston, Capt. Josiah Chapin was " authorized to discourse with, and if he think it be for the town's interest, to retain a lawyer." In 1707–8, 1708–9 and 1710, Josiah Chapin, Esq., was chosen one of a committee to instruct the Selectmen, and in 1709 he was " joined with the Selectmen in laying a way from the great road that leads to Rehoboth." In 1712, he was chairman of the Selectmen and Assessors. In 1713, he was chairman of commissioners of assessments,—committee to give instructions to the Selectmen,—committee to give instructions to committee on Cedar Swamp, and representative to the General Court. In 1714, he was chairman of commissioners for assessments. In 1715, Justice Chapin was chosen commissioner, committee on instructions to the Selectmen, and to examine the Town Treasurer's accounts. In 1716, he was commissioner for assessments. In 1717, he was committee on instructions to the Selectmen. In 1718, he was representative to the General Court, committee on instructions to the Selectmen, and moderator at town meeting. In 1720, May 10, he was chosen representative to the General Court, and June 30, he was chosen to be a representative after dissolving the Court. In 1721–22, the Selectmen met to consult with Josiah Chapin, Esq., upon sundry matters relating to the affairs of the town.

Here the record of his public life ends. It had been between sixty-three and sixty-four years since the date of his first marriage. He had become an old man, nearly ninety years of age. So intimately had he been connected with the affairs of the town, and so well did he know them, that it was natural that men younger in years and with less practical knowledge than he, should seek to avail themselves of the long and faithful experience of the veteran whose name is seen upon the records of every year from 1782 to the day when he disappears from the public eye.

Josiah Chapin must have been no ordinary man. The evidence is clear and decisive that for about forty years, although every year engaged more or less in public business, he retained the confidence of the people among whom he dwelt. Passing through various grades of office in military life from sergeant to captain, and in civil life from that of a committee to build a frame house for the minister, up to representative to the General Court, for many years the only Justice of the Peace in town, and a sort of factotum generally until like Jacob in days of old, he is found giving his parting counsels to younger and stronger men, never to the last did he experience any abatement in the confidence or respect of the community.

It is a positive luxury to investigate the days and times of one's ancestors, and to meet with a character which impresses us like that of the man whose history has been thus briefly referred to. It makes little difference in the estimate of such a man, whether the blood of prince or peasant flowed in his veins. He bore within himself the unmistakable impress of God's nobility, and his descendants of the seventh and eighth generations bow in reverence to his name and character.

Starting with the fifteen children of Josiah and connecting their fortunes, direct and collateral, with those of the various families in which their traces are to be found, we are soon afloat, and need the services of some genealogical enthusiast, to follow the courses of the family blood. No one has yet volunteered to perform this labor of love, and the genealogy of the stock of Josiah is still very incomplete. Its representatives are very numerous in eastern and central Massachusetts. They are found in every State in New England. They are scattered in the Empire State, and inhabit the rich prairies of the West. The descendants of the Rev. Stephen Chapin, D. D., formerly the distinguished President of Columbian College, in the city of Washington, are still residents of that city, while families whose names may be called legion, may trace their lineage through some one of the fifteen children of the first Justice of the Peace in the town of Mendon.

Pleasant indeed would it be if we could find that the accurate habits of Josiah Chapin had descended to those who came after him. Could we go to each one's family record, and find as full and detailed a statement of family history as we find in the handwriting of this ancestor, how easy would be the task, and how pleasant the duty, to gather up its genealogy. I know that there are those to whom this investigation is neither pleasant nor desirable; while to others it is a source of ever increasing interest. I well remember the time when, a number of years ago, I called in this city at the office of a gentleman, whose genial countenance and manly form are now before me. Upon the announcement of my name, he at once remarked to me in substance :—Do you know who your ancestors were? Being a lawyer, I told him the truth, of course. He started at once into a statement in reference to Samuel, Josiah, Japhet and Henry, which almost overwhelmed me with information as to who I was, and from whom I descended, and when, with generous warmth, he closed with the declaration that he had Chapin blood in his veins, and that he was a relative of myself, it occurred to me that it was well worth a journey from Worcester to Springfield to look upon the genial countenance and hear the enthusiastic declaration of my new found relative and friend.

Seth Chapin, a son of Josiah Chapin, became a prominent citizen of the town of Mendon, holding various offices, civil, military and religious.

Seth Chapin, Jr., a son of Seth, and a grandson of Josiah, married Abigail Adams, an aunt of John Adams. Seth Chapin, Jr., is described as "rich, very talented, one of the first men, wore a gold ring," &c.

Moses Chapin, the youngest son of Seth, Jr., by a second wife, died in the year 1802, and is still kindly remembered by some aged men as "Uncle Moses." It was said of him by his nephew, the late Adams Chapin, Esq., formerly one of the leading men in Milford, "In the year 1766, I went to Uncle Moses's and taught school in the west room, and his *latch-string was hanging out.*" His last statement must not be overlooked or forgotten. It may be a matter of minor importance that an ancestor was rich, talented, one of the first men, or wore a gold ring; but the fact that one's latch-string was hanging out, demonstrating that nearly one hundred years ago, the owner of that latch-string, even then known as " *Uncle Moses*," was the soul of hospitality and good feeling, is one which should be embalmed so far as human language can accomplish such a result, *forever*.

A fair proportion of the responsible positions of the communities where they have resided have been held by the descendants of Josiah Chapin. In Rhode Island, some of the leading business men in the State, during the last forty years, may trace their descent from the same ancestor.

In Upton, Col. Ezra Wood, who was not inaptly styled the standing representative of the town in the General Court of Massachusetts for thirty years, was a descendant on his mother's side from Josiah Chapin. In Uxbridge, the Hon. Bezaleel Taft, a member of the Senate and Council of Massachusetts, and for about twenty-five years a representative in the General Court, was a great-grandson of the same ancestor. The widely scattered members of the families of these gentlemen are living witnesses to the extent of what may be considered but a fraction of the stock of Josiah.

His descendants have done their part to acquire and protect the rights and liberties of the people. Josiah Chapin, a son of the first Josiah, was slain in Lord Russell's fight, May 20, 1693. Samuel Chapin, a grandson who settled in Wardsboro, in the State of Vermont, was a Lieutenant and had his thigh broken at the battle of Bunker Hill. Joseph Chapin, another grandson, was a Colonel

in the war of the Revolution. Eli Chapin, a great-grandson, was a member of the minute men under Capt. Timothy Bigelow, who marched from Worcester on the 19th
day of April, 1775. Benjamin Chapin, August 25, 1776,
died in the service of his country; and in this unnatural
attempt to overturn the best government upon earth,
many of the descendants of Josiah, and I doubt not many
of the descendants of the other children of the venerable
deacon are engaged in defense of the free institutions of
our beloved country.

It does not appear that the original Josiah ever reached
the dignity of Deacon; and although I have found but
two of his descendants bearing the name of Chapin who
have been clergymen, there has been a very liberal sprink-
ling of deacons in his branch of the family. Within the
memory of the speaker there have been Dea. Seth Chapin,
Dea. Nathan Chapin, Dea. Amos Chapin, Dea. Josiah
Chapin, Dea. Phineas Chapin, Dea. Samuel Chapin, Dea.
Lewis Chapin, Dea. William C. Chapin and Dea. Josiah S.
Chapin, without enumerating a multitude of deacons in
whose veins has run the family blood.

Benjamin Chapin, the ancestor of all the native born
members of the family in Worcester, and the father of
Gen. Israel Chapin, who settled in Canandaigua, N. Y.,
was a distinguished and eminent school teacher. The
names of many members of the family might be given
who have devoted themselves successfully to the same
honorable employment. (Note C.)

The leading men of Josiah's branch of the family in
former days were federalists, those of a later day were
whigs, and those now living will be left to speak for
themselves, lest the decided opinions of the speaker may
be supposed to color his judgment of his cotemporaries.

One thing he may be allowed to say, and that is, whatever may be their religious or political opinions, they have a fair share of religious and political honesty, and are not apt to ask their neighbors what they shall say, what they shall do, or how they shall vote. If I know their characteristics or the characteristics of the family generally, political ambition has not been one of their besetting sins. They have endeavored to do their duty as good citizens, and have done their part to sustain the cause of good order and good morals in the community. They have generally been firm supporters of the Christian religion and of the best Institutions in society; and when the call of duty has not been imperative, they have preferred the quiet happiness of their own firesides to the unsatisfactory pursuit of pleasure, fame or pelf. To some of the best of them may be applied, with altered phrase, the language of the poet:

> " Far from the madding crowd's ignoble strife,
> Their sober wishes *are not apt* to stray ;
> Along the cool sequestered vale of life,
> They *keep* the noiseless tenor of their way."

Friends and relatives of a common blood, I have thus imperfectly paid my respects to our common ancestor and his numerous descendants. Although embarassing, it has been a labor which has had its daily reward in the interest which I have taken in the investigation of our family history. It has rarely fallen to the lot of so many persons of one flesh and blood to assemble under circumstances so felicitous and satisfactory. We rejoice that we can look back upon a class of men so true to themselves and the spirit of the age in which they lived. We rejoice that much as we may investigate the family stock, we are almost as exclusively New England men and women as though we had been created out of the soil of this rock-bound coast, as Adam was created out of the dust of Eden. We want no higher rank of nobility, than that of descendants of the true-hearted Deacon. We are willing to feel that he and his children were the growth of New England Puritan Institutions, and we rejoice that our right to be called Yankees is so clear and indisputable. We may have come, according to one tradition, from Wales, or according to another tradition, from a source which boasts as its emblem a ducal coat of arms, but we prefer to cut loose from any foreign moorings, and under God become the architects of our own destiny. If England chooses to indulge in unjust and meddlesome conduct towards this country, we prefer not to feel one particle of responsibility for her folly, and if in her jealousy of our growth and prosperity, she shall at length involve us in a foreign war, we choose to be able to enjoy the luxury of striking home with the feeling that we are not shedding fraternal blood.

Our fathers were true to the principles and ideas of their own day. Let us not cherish the delusion that all good died with them, or that it will die with us. Meeting every issue as it comes; seeking only what is just and true, let us in the fear of God, live up to the best light within us, holding the doctrine uttered by Robinson, the great-souled Puritan Minister at Leyden, and engraved upon the heart of the world: "That the Lord has more truth yet to break forth out of His Holy Word,"—and helping, as far as in us lies, to produce " on earth peace, good-will towards men." Then when months, or years, or centuries, shall have passed away, perchance some other family gathering shall call up some recollection or tradition of this, and happy will it be for us, if our memory shall be as fondly

8

58

cherished by those who shall come after us, as we cherish to-day the memory of those whom we are proud to claim as our kindred.

NOTES TO THE ADDRESS.

NOTE A.—Upon the supposition that the foregoing address may possibly be read at some time and by somebody not familiar with the facts referred to, it may be proper to state that among the descendants of Dea. Samuel Chapin are some far-famed doctors of divinity, and some very respectable judges.

Chester W. Chapin, Esq., is President of the Western Railroad.

E. A. Chapin, Esq., is Superintendent of the Cheshire Railroad.

Marvin and Ethan S. Chapin are the proprietors and landlords of the Massasoit House.

Hon. Stephen C. Bemis is Mayor of Springfield.

All of these are descendants of Dea. Samuel Chapin.

NOTE B.—Since the foregoing was written, there has been placed in my hands an ancient document, of an extract from which the following is a copy:

"My great-great-grandfather, by my mother's side, was Samuel Chapin, Esq. Born in Dartmouth, in Old England. Come over to New England about the year 1635. Lived at Roxbury awhile, then moved to Springfield. Was a deacon of that church.

"October 29, 1779. By me, JOHN HORTON."

Having had no opportunity to test the accuracy of the statement, so far as the same relates to the birth-place of our ancestor, having found some inaccuracies in the document from which the extract is taken, and the statement having been written more than one hundred years after the death of Dea. Samuel Chapin, I prefer to give it in the form in which I find it, means having been already taken to endeavor to ascertain whether or not it can be relied upon.

NOTE C.—The number of graduates of colleges, of the name of Chapin, in this country, may be safely stated at forty.

DOCTOR HOLLAND'S POEM.

Dr. J. G. Holland, of Springfield, pronounced the following poem, written for the occasion, by invitation of the Committee of Arrangements:—

Life has a simple constitution :
We pay, or are paid, for all our deeds.
The "nub" of all things is retribution:
That is the way my Bible reads.
A bad cigar costs—a postage stamp ;
We swallow a drink, and sink a dime ;
We sow our bread where the land is damp,
And gather a harvest—after a time.
And when the land which we sow is clay—
Moist with too many a potable drop—
The operation is certain to pay,
And yield a remarkable bountiful crop.
We take, and it costs ; we give, and it pays
Measure for measure and pound for pound ;
It is compensation that fills the days,
And makes the grand old world go round.
Good has its price to the weak and the strong,
To farmer and artisan, priest and scribe,
And I am compelled to sing you a song
For marrying into the Chapin tribe.

And what shall the song be ? So it be brief,
It matters but little in any way;
We have read the grave side of the family leaf,
So, just for a change, let us turn to the gay.

Did it ever occur to any of you
That this old Deacon Chapin, of whom we read,
And about whom we raise such a hullabaloo,
Was a very remarkable person indeed?

cl 'Tis a philosophical axiom, I'm told,
As solid as gold and very old—
As old almost as the mountains—
That streams, no matter how large they may be,
In all their beautiful paths to the sea
Never rise above their fountains.
Well, look at this river of family life :—
Thousands of hearts pumping blood by the cask—
And when you are satisfied, let me ask
What you think of the source—and the source's wife.
No matter how high the particular shelves
On which we see fit to exhibit ourselves,
No matter how many we reckon,
There is nothing (short of a water-ram)
That can put us above old Deacon Sam
And the woman who fried his bacon.

Why, when the old town was settled at first,
They were all afraid that its bounds would burst.
The men were so broad in the back and so tall,
That they did'nt want neighbors within a mile ;
And their fruitful wives, neither short nor small,
Multiplied wondrously all the while.
So they fixed, as you know, at a couple of score
The number of families; then they swore
In the Puritan way, with a "snum" and a "snore,"
That the broad old town would'nt hold any more,—
Proving beyond any question that then,
The valley was peopled by rather large men.

Why, I have no doubt that our Deacon Sam
Could have walked from Springfield to Wilbraham
In fifteen minutes; or brought in from Skipmuck,
With as little fatigue as I could a chipmuck,
A buck or a bear so uncommonly thrifty
As to weigh by the steelyards five hundred and fifty.
And we know that though timber was plenty and cheap
And the river in places decidedly deep,
Not a bridge was put up till the old people faded,
And I take it for granted that all of them waded.
I fancy I see them step forth from their houses,
Rolling up in their progress the legs of their " trowses,"
With their shoes in one hand and their wives in the other,
(They hadn't on any long dresses to bother)
And somehow get over the charming Connecticut
Without scaring a woman, or dipping a petticoat.

Now if streams send no ripple above their source,
(And they never do as a matter of course,)

Suppose you endeavor to tell me whether
This great-great-grand—this immensely grand father
Didn't know more than all of us put together.
What *have* we—what *are* we—in body or soul
That is not from the Deacon, in part or in whole?
You know that the Reverend Henry Ward Beecher
Is considered a very remarkable preacher
And, in many respects, quite a marvelous creature.
You have heard him within his own pulpit on Sunday,
And lecturing somewhere or other on Monday,
And been puzzled, perhaps, with tormenting conjecture,
As to which was the sermon and which was the lecture;
But sermon and lecture have set you aflame
With the fire of an eloquence always the same.
Week in and week out, and year following year,
You have heard his great voice ringing roundly and clear,
Proclaiming the right and denouncing the wrong,
And defending the weak in their strife with the strong;
Yet gentle and playful and dainty and airy,
As the flight of a fawn or the foot of a fairy,
His fancy has danced evermore with the Hours,
To the tune of the birds and the time of the flowers.
And you have looked on but to wonder and wonder
Where this eloquent son of an old son of thunder
Obtained the rare gifts that have crowned him the chief
Of the men who proclaim the old fashioned belief.
You have thought it was genius, or something still higher,
That filled him so full of electrical fire.
You have thought that the gods had descended, perchance,
And massing their forces had made an advance,
Driving in Nature's pickets, or shooting them dead,
And taking position on Beecher's broad head.
You are widely mistaken; the gods do not aid him;
He is just what the blood of the Chapins has made him.

There is Hitchcock, a great theological doctor,
A preacher of power and a notable proctor;
With genius unquestioned, a name above scandal,
And doctrines all orthodox "up to the handle;"
Who never has dreamed, I may venture to say,
That he came by his creed in the natural way,
And that all that he carries of power and persuasion
Is old Deacon Chapin, and not inspiration.

There is Chapin, the preacher, whose avoirdupois
Will equal the best of the old Chapin boys;
With a heart so immense and a form so abounding,
A brain so astute and a voice so astounding,

That he drops in his creed the old-fashioned rehearsal,
And spreads into schemes that are quite universal.
You're aware, I presume, that he's not in this quarter,
But is gone from the country far over the water;
And gone for his health (so the editors say,)
Which is delicate, quite in the clerical way;—
A way, as they tell me, which comes from endeavors
To raise heavy weights with incompetent levers,
Or from fruitless and feverish strife to command
A very good game with a very bad hand,
Or—to leave in illustrative figures no gap at all—
From doing large business on very small capital.
Now the doctor has brains enough, every one knows,
But they're regular Orthodox brains, I suppose.
They came from the Deacon, whose scheme of salvation
Was yoked with a parallel scheme of damnation.
So our friend in proclaiming his easier creed,
Though with muscle and mind quite enough for his need,
Simply went against nature—the family blood—
Which has always been Orthodox back to the flood;
And in preaching the worlds absolute absolution
Broke the Chapin all down in a strong constitution.
Ah, sad is the Doctor, and sorry are we
That he comes not to-day to the family tree;
But we send him our greeting far over the sea,
And we breath our desire, which we lift into prayer,
That the God of his father may be with him there.

But the blood hasn't all run to Davids and Daniels,
Or followed alone theological channels.
It flourishes everywhere—favors all classes,
And thrives in those aggregates known as the "masses."
The Shakers and Quakers and photograph-takers,
The butchers and bakers and candlestick-makers,
Will furnish you Chapins by acres and acres.
It presides o'er a college in distant Beloit,
And feeds half the world at the old Massasoit.
Its decisions in probate are Equity's pride,
Though doubtless they lean to Humanity's side;—
That is, although just, they are apt to consider
The orphans, and lean to the side of "the widder."
Of these matters of probate in Worcester, I know not,
And therefore of matters in Worcester, I blow not;
But in Hampden, I know how these things were decided,
When the old court existed and Morris presided.
It carries the mail over half of the mail-roads,
It manages stages and steamers and railroads;
It is engineer, financier, captain, constructor;
It is president, treasurer, clerk and conductor;

In short, without further expansion to trouble you,
It is just what we find in our own "Chester W."
Though it occupies not the state minister's chair,
It sleeps in his bed and partakes of his fare,
And bears him his boys, as you're doubtless aware;
And the Foote of the family, you may remember,
Is the head of the national Senate chamber.

But enough. I have told you sufficient to prove
That the family runs in a very broad groove,
And that when we come down to the matter of blood,
It is "thicker than water," and—thinner than mud.

The name is an old one, and dates from the ark
When Ararat's summit was high-water mark;
When all the distinctions of caste lay between
The chap that was out and the chap that was in.
The chap out was drowned in the great inundation,
And the Chap in was saved for the new generation.
Our Chapin was Japhet, a name that we find
Has been much in the family, time out of mind;
The identical man, as you doubtless discover,
Who was "one of three brothers," that left and "came over."

But enough in this key; let us change to a sweeter
And close in a much more particular meter.

I have a fancy that the forms of some we do not see
Are gathered with us as we throng around the household tree;
They fill the windows and the doors, the galleries and aisles,
Silent in dignified delight, and radiant with smiles.

Brave William Pynchon comes to-day (whom men and ghosts revere,)
With Major John, "the Worshipful," and stands beside me here;
And old John Woodcock, free at last, and wearing not a scar
Of the old chains, bows reverently, and worships from afar.

Oh up and down these pleasant aisles, on Holyoke's manly arm—
The woman to an angel grown—still crowned by woman's charm—
Fair Mary Pynchon walks unseen, the maid, the peerless wife,
The beauty and the glory of the old plantation life.

Moxon, the pastor bends with eyes abashed before the gaze
Of those who knew and feared him in his superstitious days,
When witchcraft's ban lay on his home where Christian peace should be
Till racked by hellish fantasies he fled across the sea.

All these and other forms there are that meet with us to-day,
The little ones—the nameless ones—who lived and passed away
While yet the beaver built his boom, the salmon swam the floods,
And red deer ranged unchallenged still the circling solitudes.

But happier than all the throng—tho central shade, indeed—
Is the benignant saint whose children listen while I read.
Lord of tho festival is he, whose fruitful loins have given
Ten thousand forms to walk the earth, ten thousand souls to heaven.

Then let us fancy—nay, believe—that from their golden home,
The father of our flock and all of kindred birth are come,
And realize that in the joy which thrills each kindling mind,
They bring us precious bounty from the heaven they left behind.

THE PRESIDENT.—I suppose you are all aware that it was expected that Rev. Henry Ward Beecher would be here to-day, but I have a letter from him, in which he states that it is impossible for him to come, on account of ill health. In the letter, he acknowledges the blood, and I will read it to you.

LETTER FROM REV. HENRY WARD BEECHER.

BROOKLYN, September 16, 1862.

MY DEAR SIR:—When I promised to be present at the great family Chapin Meeting, I expected to have been free from an opthalmic catarrh, whose annual visitations usually are ended by the middle of September. But I find myself unable to bear the dust and cinders of the cars, and shall be obliged to deny myself the great pleasure which I had anticipated. All the Chapin blood in my veins rises up against this decision. But in vain. I can only hope a better fortune for the multitude of that name or blood, who propose to attend the celebration. I shall keep the day at home in my own way.

Upon looking back to the records, I find that a granddaughter of Dea. Chapin, in 1672, or 1676, for records differ, married Nathaniel Foote, the ancestor of my own mother, Roxana Foote. The stream was already mixed with Bliss blood when it began to move down the Foote channels. On its way to me, the Chapin blood received important additions from many sources, until now I am at a loss to know how many sorts of blood I have racing through my veins. There are the Blisses, the Jays, the Wellmans, the Wards and the Hubbards, and many others, on one side. The Roberts, the Stowes, the Sperrys, the Howleys, the Lymans on the other, together with drops from other named veins. I have therefore decreed to all the blood of all the families in my veins, that to-morrow, Wednesday, September 17, they do reverence to the Chapin blood. To-morrow morning that blood takes the lead in my veins, and for the day, is honored head and president of all the rest. All the Howleys, Sperrys, Stowes, Roberts, and Lymans, within me, shall rise up and call the Chapin blessed! The Bliss-Jay-Wellman-Ward-Hubbard congregation shall rejoice in the presence of the honored Chapin in me, and do it reverence!

To all, in every State, (except the State of Rebellion,) that carry the Chapin blood, with or without the family name, I extend a greeting! Kindred all!—hitherto unknown, and now unseen, I give you heart and hand, as a true man, faithful ever to kith and kin.

H. W. Beecher.

SPEECH OF REV. A. L. CHAPIN, D. D.

Rev. A. L. Chapin, D. D., President of Beloit College, Wisconsin, was then introduced, and spoke as follows :—

Fathers and Brothers,—Kindred all,—I feel that this is a joyous, blessed, and, as just sketched in the closing lines of the poem to which we have listened, a solemn presence. How to meet the demands of the occasion, in the very few moments that can be allotted to me, in justice to others, is an embarrassing question; but as the orator of the day claimed for himself, and for all of us who might follow him, the utmost freedom to speak without fear of criticism, without regard to anything but the promptings of our own hearts, now full and warm, as we feel the tie of this kindred assembly, I will sketch briefly the line of thought that has been passing through my brain as I have been listening to the statements of the orator and to the beautiful strains of the poet.

As descendants of that good old Christian Deacon, it becomes us here, with lively gratitude to God, to recognize the connection between the character of our ancestor and the results of his life, as it appeared in his own action and in his descendants. In serious rather than sportive strain let us on this festive occasion, joyfully acknowledge God's fidelity to His covenant, which has stood in ages past, and shall stand for ages to come. I need but run over those which seem to me, according to the scriptures in which our father trusted and which we all accept as the record of divine inspiration, the blessings promised to the pious patriarch.

The promise speaks first of a multiplied offspring : the proof of its

9

fulfillment in this respect we have before us. God has blessed our ancestor, according to his covenant, and rewarded his piety with that which the Psalmist says is his reward, " Children are the heritage of the Lord, and the fruit of the womb is his reward." One of our number here present, taking such partial data as could be gathered from the book of Genealogy recently prepared, has guessed,—(of course, it can be but a guess,) with some good reason, I think, that down to the present time, there must have proceeded from the loins of that father whom we honor to-day, at least thirty thousand souls. It seems to me there is in all our land, hardly another instance of such a fruitful family as this which gathers here to-day to honor their father, and the God of their father.

Another blessing promised to the pious, faithful patriarch and to his descendants is long life. That, as was clearly shown in the remarks of our orator, has been a peculiar characteristic of this Chapin line. I stood, yesterday afternoon for the first time, in the old burial ground of Chicopee, and marked the ages of those of the Chapin family who lie buried there—outnumbering every other name in the ground—and their length of days is remarkable, ranging all along, with only here and there an exception of one dying in infancy or childhood, from seventy to ninety years. The heads silvered with age before me tell us that God continues that blessing still in our line.

Another of the blessings promised to the faithful, godly soul and to his posterity is that they shall be cared for in this world—provided for with all things needful for their comfort. The Psalmist says, " I have been young, and now am old ; yet have I not seen the righteous forsaken, nor his seed begging bread." Has not that promise been fulfilled peculiarly in this our Chapin line ? Very rarely indeed do we meet with one who would be accounted very rich in this world's goods. Now and then there is one who stands thus marked among us. But never have I met with a man of the name—though I have met with them in most of the Northern States—who could say of himself, or of any of the name he ever knew, that he suffered from poverty. The prayer of Agur, " Give me neither poverty nor riches," seems to have been wonderfully answered in our line, according to the precious promise of God.

The preservation of the godly character in the line of the pious ancestor is another of the promised blessings. How that has been illustrated in the history of our family, the record of ministers and deacons (partial, very partial) which has been given us to-day, conclusively shows. They would be counted by tens and scores, rather than by units, if we could gather all those who have held places of responsibility and influence in the church of Christ. We stand, I think, as a family, marked peculiarly by the preservation among us of an abiding faith in this Gospel of our Lord and Savior Jesus Christ; and mostly, too, in the line of that peculiar faith which was professed by our ancestor. The divergence here and there, on the part of individuals, has not carried them, we believe, beyond the range of that which is for themselves—however any of us may think of the tendency of their doctrines—a real and saving faith in the Lord Jesus Christ. The number marked by sincere piety, it seems to me, is remarkable, for a single family, the descendants of one patriarch.

Now, we must recognize all this as running in the line of that blessed covenant of grace, proclaimed in ages past, to the old patriarch Abraham, to which our God has been faithful in the ages and generations that have followed. As I stand here in this sacred place, on the spot, if I am not mistaken, where the old church stood in which the Deacon—our common father worshiped, I love to let my heart, swell with the beautiful thought which closed the poem —that there is an unseen presence here, that there are hundreds and thousands, kindred to us in flesh and blood when they lived on earth, kindred to us, as we believe, also in the faith of a blessed Savior, who are looking down upon us, rejoicing with us, and glorifying God as we cannot, for that which they are permitted here to-day to see of His faithfulness to His covenant, and of the blessings He has given to the pious patriarch and his descendants. Our hearts are drawn together in sweet and kindly sympathy towards each other; and the same sympathy,—for it is God's design, as I understand it, that we, while sojourning here in the flesh, shall be linked to things invisible and immortal by those ties of sympathy begun on earth,—those ties of sympathy run on and unite us here to-day with that great multitude, our immediate kindred after the flesh, who have gone to join the great assembly and church of the first-born, whose names are written in heaven. O! it is a blessed thing to stand here and feel this. It is a blessed thing to have our hearts and souls lifted up to God in devout gratitude for it.

And then, how there comes down upon us another thought— its counterpart—the responsibility that attaches to those whom

God has peculiarly blessed! "Of them to whom much has been given, much will be required." While we rejoice and lift up our souls in devout gratitude at the one line of thought which I have just suggested, let us also open our souls to feel the influence of the other. There is binding upon us, by all the regard we cherish for our ancestors who have gone from earth, by all the regard which we have for that God who is faithful to his covenants, the obligation to perpetuate their spirit of heroism and their Christian character among those who shall come after us, that they may be true, in their day and generation, to the great principles of duty which their ancestors professed, and at last join them in that other nobler and purer worship above.

I had thought to illustrate some points of this course of remark by quotations from a letter which came into my hands a short time ago, written some twenty years since by Dr. Calvin Chapin, of Rocky Hill, Conn. I felt that it might perhaps be allowed me, standing as I do in this presence to-day, the only lineal representative, certainly who bears the Chapin name, of the good old Deacon Edward, so cherished, and so honored in the history of the neighboring village of Chicopee, to read a worthy son's sketch of his father's worth ; but I know there are others waiting to speak, and that you are waiting to hear them, and I will not do so. But there is one thought to which in closing, I will call attention—as it seems to me one which we may well take up and carry hence with us. Speaking of one of his brothers who died, he desires to know whether he was true to the characteristics of his family; that is, to quote his words, " whether he so lived by grace as to trust that Christ owned him as His brother, and that consequently he, without presumption, might call Christ his brother," and adds, " surely, I cannot even imagine any other favor in the universe equal to that." May we not all of us take the word which comes from that departed saint, and endeavor to cherish a faith which shall bring us, as a whole family, up into relationship to that crucified Savior, so that He shall be our brother, and we His. And so, moving on towards the consummation of His gracious designs for a lost world, may we be gathered, with all His redeemed ones, into the true and final home of all the kindred of Christ, owned and accepted in the better world.

The following quartette, entitled " Home Again," was then sung with exquisite taste and feeling:

Home again, home again,
From a foreign shore,
And oh it fills my soul with joy,
To meet my friends once more;
Here I dropped the parting tear
To cross the ocean's foam,
But now I'm once again with those,
Who kindly greet me home;
Home again, home again,
From a foreign shore,
And oh it fills my soul with joy,
To meet my friends once more.

Happy hearts, happy hearts,
With mine have laughed in glee;
But oh! the friends I loved in youth,
Seem happier to me;
And if my guide should be the fate,
Which bids me longer roam;
But death alone can break the tie;
That binds my heart to home.
Home again, &c.

Music sweet, music soft,
Lingers round the place,
And oh! I feel the childhood charm,
That time cannot efface,
Then give me but my homestead roof,
I'll ask no palace dome;
For I can live a happy life,
With those I love at home,
Home again, &c.

SPEECH OF REV. ROSWELL D. HITCHCOCK, D. D.

Rev. ROSWELL D. HITCHCOCK, D. D., of New York, was then introduced by the President, and spoke as follows:—

During the summer and early autumn of the year of our Lord 1666, while Old England, now so lamb-like, was up to her eyes in war, —fighting France, fighting Denmark, fighting the Dutch,—here in Springfield, on the banks of the Connecticut, in New England, John Hitchcock was "courting" Deacon Chapin's daughter Hannah. Some of his descendants, I dare say, are now "paying their addresses" to the ladies, but John "courted." (Laughter). What strategy he employed, if any, I have not been told; but my private opinion is, that he employed no strategy at all, but like our modern Ulysees, whose other name is Grant, marched straight upon the works. At any rate, he carried the day. Hannah said "yes" (laughter); the good Deacon and his wife said "yes;" and on the 27th day of September, the young people were married. It was that wedding which brought me here to-day. (Laughter and applause.) Somebody (some one of the Chapins, it is altogether likely) has remarked, that " it takes a wise son to know his own father." It seems to me that we have heard enough to-day to persuade us, if we needed persuasion, that no man is wise enough to know all his own relations. I have been a Chapin, it seems, all my life, and knew nothing about it until Judge Morris told me, two or three months ago. I have had Chapins to teach in the college and in the seminary; for some years, I have been side by side with the D. D. so famous for his heterodoxy, and not much farther off from that other *would not be* D. D., quite as remarkable for his orthodoxy, and did not know them for cousins. (Laughter.) And then I have eaten, and paid for, innumerable excellent dinners and suppers at the Massasoit House, the landlords evidently not knowing that I was a city cousin. (Renewed laughter). It is a very curious matter, this of blood. There is no end to it. Human blood is like the sea; every drop tinges and is tinged of all. Human relationship is so near to being universal, that our arithmetic cannot demonstrate that it is not. I remember being amused, a few years ago, on taking up a book on the chances, prepared by my old friend, Professor Fowler, of Amherst, to find him tracing back the blood of President Chauncey, of Cambridge, to the loins of William the Conqueror,— to which, I suppose, our orator would not object, since the Normans whipped the old English and took their country,—and back to the loins of Charlemange. Very likely it is so. If I were going into any such investigation as that, I think I should choose not to stop short of the venerable ancestor referred to by the poet, Japhet;— not Japhet Chapin, but Noah's Japhet. As matters now stand, I certainly would not like to be counted in among the descendants of Ham (laughter), since the President's proposition in regard to colonization has not yet been definitely acted upon, and I do not know where they are going; and as the race of Shem seems to be somewhat effete, I choose to stand, and stand jubilant, among the Japhetites. But then, I sympathize very strongly with the orator, in disclaiming any desire to go beyond our New England beginnings. I have no itching in my finger tips for any proved relationships beyond the sea. I have a strong sympathy with that man who burned the books at Alexandria, saying—"If they echo the Koran, they are needless; if they dispute the Koran, they are a nuisance." Most things which happened before New England was settled may be put out of the account, and not much has happened since New England was settled of much account, but what has happened in New England, or through New England influences.

Now, Chapins, male and female, where do we stand? What do we represent? It seems to me that the great inspirator of our modern civilization, Martin Luther, struck the key-note of it when he said of himself—"I am a peasant's son ; my father, my grandfather, my ancestors, were all peasants, and my father was a God-fearing man." The American population, as you know right well, occupy the continent in three great lobes : one-third, English middle-interest —New England ; one-third, Southern English—decayed gentlemen, for the most part—goldsmiths broken in their business in London (not to be too severe upon our Southern brethren) ; the other third is made up of all the European races. This New England third has been the organizing third ; like the Saxon in our language, it has been the syntax of our history. The Northern third, New England, has thought, and felt, and spoken, has given the word of command, has marched and conquered, and must march and conquer to the end. All that we need to make out in regard to ourselves, it seems to me, are these two things—the only patent of nobility that I crave : first, that our ancestors feared God with a fear so imperial as to crush all other fears ; and, secondly, that they ate bread moistened by the sweat of no other brows but their own. (Applause.) This is the glory of the Chapins ; it is the glory of New England ; and with these two ideas, Providence has now marshalled us in the serried lines of battle. The conflict is on us, and the victory is sure. New England is more than these six little States. New England belts the continent, for her blood has washed westward to the Pacific, and her thought has organized the life of the continent and electrified its activity ; and the conflict is now between this New England English third, and the Virginia English third. The fear of God—I will not say it has been disclaimed on the Southern side, for I know they pray, and I have some sympathy with that good Baptist deacon who said the other day—mention having been made of the fact that there were praying men on both sides the line—" Yes, and they have got the Lord in a pretty tight place." (Laughter.) I will not say that the fear of God has been disclaimed at the South—I will not vaunt our piety ; but I am proud to say, the bread we eat is not moistened by the sweat of sad black brows. We eat what we have earned in the fear of God; and in the last analysis, here lies the edge of battle—the civilization of the Christian North, which eats its honestly-earned bread in the fear of God, and the civilization (if you will) of the equally Christian South, which offends against all the laws of history by eating that for which it hath not toiled. The dust of the conflict and its din are upon us and about us. I rejoice that this new recruiting now in progress is going on in doubt and disaster. I will not criticise the President or his Cabinet ; I will not criticise our generals ; I only say, allow the worst or believe the best, in either case, it is the people who must save themselves ; and if we understand our calling, fear God, eat bread honestly earned,—accepting our mission, and accepting the conflict which our mission has imposed upon us in this Thermopylæ of the centuries,—we cannot fail, until God himself shall abdicate. (Applause).

SPEECH OF HON. SOLOMON FOOT.

Hon. SOLOMON FOOT, of Vermont, was next introduced, and cordially welcomed. He spoke as follows:—

Ladies and Gentlemen,—Fellow Citizens,—Friends and Kindred, —You have my sincere thanks for this generous and complimentary salutation. I have to express my acknowledgments, with many thanks, to your Committee of Invitation, and to your Committee of Arrangements, for the opportunity of appearing here as one of your number on this most agreeable and interesting occasion. It is no little matter of personal gratification and pride to me, let me assure you, that I can claim kindred with that large and highly respectable number of our fellow-citizens, scattered throughout this and throughout many other of the States of this Union, the descendants of that early pioneer, and one of the most prominent and worthy of the pioneers in the settlement of this country, Deacon Samuel Chapin—the reputed, and I have no doubt the real, common ancestor of all who bear that name in the United States, as well as of numerous branches bearing other names, through the female line of descent. I stand in your midst, and I am proud to say it, although personally a stranger to most of you, a representative—would that it were a more worthy representative!—of one of these branches, as a direct lineal descendant of Deacon Samuel Chapin, through his eldest and most excellent daughter, Catharine Chapin. I therefore tender to you all, as I am most happy to accept from you all, my salutations, my greetings. I tender to you all, as I am proud to accept from you all, the right hand of fellowship, the right hand of friendship and of kindred.

It may not be altogether uninteresting, it may not be entirely out of place, on an occasion like this, even though it may savor somewhat of personal vanity or arrogance, to trace out, very briefly, my own family lineage, so far forth, at least, as to establish my own claim to this fellowship. We have already learned, from the instructive and admirable address to which we have all listened, I have no doubt, with great interest and high satisfaction, that as early as 1648, on the 20th day of November of that year, Catherine Chapin was married to Nathaniel Bliss, of this town of Springfield, where Deacon Chapin and his family then resided. In 1672, according to my recollection (and it is only three days since, happening to be in Boston, I passed a few hours in the rooms of the "Genealogical Register," investigating this matter)—in 1672, Margaret Bliss, the daughter of Nathaniel and Catherine Bliss, was married to Quartermaster Foot, then of Hatfield, afterwards of Springfield, and finally of Weathersfield, where he died in 1703, at the age of fifty-five years. And now I can say, with my newly-discovered cousin, Rev. Dr. Hitchcock, that it was that wedding between Margaret Bliss and Nathaniel Foot, in 1672, which has brought me here to-day. (Laughter). Nathaniel Foot was a man of some considerable eminence, distinction and influence—a leading man in his day. His widow, the grand-daughter of Samuel Chapin, survived her husband, Nathaniel Foot, many years, and died in 1745, in Winchester, Conn., at the advanced age of ninety-five years. These were the grand-parents of my paternal grandfather. They reared a family of nine children—five sons and four daughters; the third son being Josiah, who was my great-grandfather, and the great-grandson of Samuel Chapin. Josiah appears to have been a common Christian name in both families. It was the name, as we have already learned, of one of the sons of Dea. Samuel Chapin. Josiah Foot, my great-grandfather, in 1712, was married to Sarah Welles, of Winchester, Conn., where he died in 1778, at the age of ninety years; his wife having deceased in 1766, at the age of seventy-three years. They reared a family of eleven children—seven sons and four daughters. The second son, Jonathan Foot, was my grandfather; and this Jonathan Foot, in 1749, was married to Sarah Fenner, of the distinguished family of that name in Providence, R. I. They resided for some few years in the town of Saybrook, Conn., when they removed to the town of Winchester, in the same State, where they resided some twenty-three years, when my grandfather removed to the town of Lee, in the county of Berkshire, where he died in 1803, at the age of eighty-eight—his wife having died some twelve years before him, at the age of sixty-one years. They reared a family of eight children—five sons and three daughters. The youngest son, and the youngest child of the family, Solomon Foot, was my father, who removed to the State of Vermont, (or, more properly speaking, went, for he had no family at that time,) to the town of Granby, where he settled as a practising physician; but in 1804, he removed from thence to West Rutland, where he died in 1811, at the age of forty-three years, leaving a widow and four children, myself being one of the number.

Thus it appears that Dea. Samuel Chapin, the progenitor of all the race of that name, and those connected with it, was the great-grandfather of my great-grandfather, Josiah Foot, who was the son of Margaret Bliss, who was the daughter of Nathaniel Bliss and the daughter of Catharine Bliss, who was the daughter of Catharine Chapin, who was the daughter of Dea. Samuel Chapin.

I have thus given you, in brief, my family lineage, and exhibited my title and right to the privilege of a seat here on this occasion. This is the record, I will say, of my family nobility, if you please; the evidence of my claim to recognition in this family circle. Perhaps I ought to suggest, for the benefit of my numerous Chapin relatives, that although our connection is direct, and clear, and certain, nevertheless it is so remote, that it will not do them much harm, nor me, I fear, much good. Nevertheless, I recognize and am proud of that relationship. I stand here to-day, and we all stand here to-day, at the end of a long line, running through many generations, and more than two centuries, yet in direct descent from one common American ancestor, whose descendants to-day are to be counted by multiplied thous???, who are dispersed over a vast extent of country and thro??? ??y States, and occupied in the various departments ??? ???? ?? public and private life; many of them filling wit? hi?? ?? ??? and credit the various liberal professions; some of them occupying high places of honor and of trust; and all of them, I am happy to believe, maintaining an honorable position among their fellow-men, not unworthy of the name, the character and the memory of their distinguished and revered progenitor.

I am very well aware that the recital of these details of personal history and of family genealogy is of little importance, and imparts little of interest to any outside of the immediate family circle of connections and descendants; but I remember (and that is my apology, if any apology is needed) that to such chiefly, if not to such only, I am now addressing myself; and I have made these references, thus far, for the further reason, that as yet we have but a very imperfect knowledge of the personal history and genealogies of the many branches of the descendants of Samuel Chapin, except that one to which I belong. I have already learned something of them sitting here to-day and listening to the addresses which have been made, and I hope to learn much more from this family gathering. And therein lies, let me say, the great value of the great interest of these family reunions, aside from the immediate social pleasure which we derive from them. Reverence for one's ancestry, special personal regard for, and special personal interest in, those who are allied to us, even though but in a remote degree, by the tie of consanguinity, is a sentiment both natural and universal among mankind. It is a sentiment, too, as pure, as virtuous, and as noble, as it is natural and universal. It is a sentiment to be cherished, to be cultivated, to be strengthened. It is, indeed, the source of very much of our domestic and social happiness. It is an unselfish, a generous, and a liberalizing sentiment, if I may use the expression. It is a sentiment of benevolence, it is a sentiment of refinement, it is a sentiment of high moral culture. It is a sentiment, then, I repeat, which is to be cherished and cultivated and strengthened; and to this end let me say, let these family gatherings, these family festivities, these family reunions, be repeated and multiplied. Let the old folks and the young folks, the grandfathers and the grandmothers, the fathers and the sons, the mothers and the daughters, the brothers and the sisters, the neighbors and those who are afar off, those of low degree and those of high degree, those of humble and those of exalted rank,—let all who are in any wise allied by the blood of a common origin, come together on more frequent occasions to these family festivals, as members of a common household and as members of a common brotherhood. The influence of it will be for good. It begets a mutual interest, a mutual respect, a mutual sympathy, a mutual attachment; and these will make us better neighbors, better friends, better citizens, better patriots, too. Hence I approve and most earnestly commend the motive and the object of this family gathering here to-day. We shall all be benefited by it, we shall all be instructed by it, as well as highly entertained by it. We shall know one another better and respect one another much more than before. We shall know more of the kindred of our household, and regard them with more favor, than we have been accustomed to do before. We shall feel a deeper sympathy for and take a deeper interest in the prosperity and well-being of one another, than we have done before. I am sure, that whenever or wherever I may meet a Chapin, or one claiming descent from a Chapin, I shall almost instinctively, as it were, recognize in that person, stranger though he may be, one of my own family and my own household,—a family connection, a family relative, and shall feel myself as in the companionship of one of my own household, of one of my own blood.

I need not say to you how I shall carry away with me, how I shall carry with me through life, many agreeable recollections, many pleasing memories, of this family gathering to-day; how it has enabled me to form many new, most agreeable, and (I trust they will prove) lasting acquaintances; how it has enabled me to become better acquainted with those of my own kindred and household than before, and therefore to regard them with a higher esteem, and to cherish for them a warmer attachment and sympathy than before.

But, my friends, I must detain you no longer. I came here without preparation, without the purpose or expectation of making any speech. I have been absent from home until yesterday, and was in doubt, in exceeding doubt, whether I should be able to gratify my desire to be here at all, until within an hour of my departure. I will therefore close these few desultory remarks with this general sentiment, which I doubt not, will receive your united and unanimous approval. I give you—The sons and daughters of the descendants of Deacon Samuel Chapin, wherever they may be, or whoever they may be, the older they grow, and the bigger they grow, and the more numerous they grow, the more welcome may they always be to this old family homestead. (Applause).

SPEECH OF JUDGE MORRIS.

Judge MORRIS, of Springfield, was then introduced, and spoke as follows :—

Ladies and Gentlemen :—In the course of the Mayor's address, he referred briefly to this city. It is indeed a city of which I am proud; it is the city where were born my maternal ancestors; the city in which I have lived more than sixty years, though born in the immediate neighborhood; the city where repose the ashes of my wife, and I trust mine also will repose within its bosom. The hands of the clock admonish me that I must speak to you only a few words. Like our friend from Vermont, I descend from the oldest daughter of Dea. Samuel Chapin, Catherine Chapin, who married, in 1672, I believe, Nathaniel Bliss. I spring from her oldest child, Samuel Bliss, who was born "a little before sunrise," as the record says, and was sent off to the minister to be baptized the same day—his parents probably expecting that his life would be brief. If so, he disappointed them, for he lived till he was almost 102. (Laughter). I can say more of my descent from Catherine Chapin. You have heard to-day that she had three husbands. By the first, she had four children, by the second four, and by the third, four, though one of them died very young. I am also descended from her through her tenth child, by Samuel Marshfield, so that I have a double relationship to Catherine Chapin, and have a right, therefore, to claim kindred with all who claim to belong to the Chapin family. And it is a noble family. I look round and see the faces before me, and I see they are all honest faces; they give evidence that they spring from an honest root.

There are one or two coincidences to which I will allude, and then I will relieve your patience entirely. Samuel Chapin, the great ancestor of us all, died in the month of November, 1675. All of you who are acquainted with the history of those times know that the year 1675 was a bloody year in the annals of New England. The people of New England were then engaged in a war more terrific than any other in which she has ever, before or since, been engaged, and in which her sons literally poured out their blood like water. It was estimated that every sixth family in New England was in mourning; that every sixth house was the habitation of some one who had lost a friend in that bloody war. It was indeed a terrific war; and we are met here to-day when not only New England, but the whole country, is engaged in war. I need not enlarge upon it. You all feel the importance of the result. I have no doubt that it will be propitious.

There is another coincidence, which I wish merely to name. This day is the 17th day of September. Seventy-five years ago to-day, George Washington placed his hand to the Constitution of the United States. (Applause). We are met here, my friends, at this Chapin gathering, on the anniversary of the formation of the Constitution of the United States,—a Constitution against which evil men have risen up—a Constitution in support of which we are sending our sons to the battle-field; and I trust there is no father or mother present who is not willing to give up his or her son to sustain the government under which we live, and the Constitution under the influence of which this country has flourished as no other country of the world has ever flourished.

Now, my friends, you know where I stand. I stand here a descendant of Samuel Chapin. I stand here in support of the Constitution of the United States of America, which I have again and again sworn to defend; and were I not so old—the frosts of eighty winters being upon my forehead—I would ask you, young men, to go with me, and jeopard your lives in defence of your country. (Applause).

The first and last verses of "America" were then sung, and the exercises at the church terminated, the large company forming in procession, (with the members of the city government and invited guests,) and, preceded by the Springfield band, marching around the square to the City Hall, Col. Harvey Chapin acting as chief marshal.

COLLATION AT THE CITY HALL.

The company having entered the Hall, a blessing was invoked by Rev. Mr. Buckingham, in these words:—

Father of Mercies, Thou art the Author of our being, and the Giver of all these gifts. We thank Thee for these relationships of life—for our kindred and our friends. We thank Thee for the memory of our fathers, and for the favor with which Thou hast blessed them in the walks and offices of life. May Thy blessing rest now upon this assembly, and upon all who are connected with them by the ties of kindred or friendship. May all who bear this family name, wherever they go, be faithful to the principles which they have inherited; may they be useful and a blessing to the world, and heirs of heaven. And may Thy spirit be with us now and sanctify this occasion; which we humbly ask, for Christ's sake—Amen.

The family were then invited, by His Honor the Mayor, to partake of the viands and fruits which were temptingly set forth on the bountifully-spread tables, and an hour or more was pleasantly spent by the company about the "family board," the band meanwhile enlivening the scene with several popular airs.

When the repast was finished, the President called the assembly to order, and, on motion of Judge Morris, of Springfield, the thanks of the family were unanimously tendered to Hon. Henry Chapin, of Worcester, and Dr. J. G. Holland, of Springfield, for the Address and Poem delivered by them respectively, and they were requested to furnish copies of the same to the Committee of Arrangements for publication.

Letters were then read from several absent members of the family, and, after considerable time spent in fraternal greetings, in the renewal of old acquaintanceships and the formation of new, the company separated, and the public ceremonies of the day were brought to a close.

LETTERS.

The following are a few of the letters received from persons not able to attend the meeting :

LETTER FROM HON. WILLIAM H. SEWARD.

DEPARTMENT OF STATE, ⎱
WASHINGTON, 8th August, 1862. ⎰

S. C. BEMIS, Esq., Springfield, Mass.—Sir: Your invitation to me to attend a gathering of the Chapin family at Springfield, on the 17th of September next, has been received, together with the pleasing information that, further, it is my marriage that has brought me within that extensive and respected circle. I regret that I shall be unable to participate in the proposed festivities, which I regard as not merely innocent but also beneficent in their influences. Just now I am engaged in the endeavor to preserve the integrity of a family even larger than that over which you preside, and one upon whose salvation the safety and happiness of all domestic relations in the country depend, and am therefore unable to be absent from the Capital.
I am, sir, your obedient servant,

WILLIAM H. SEWARD.

LETTER FROM REV. D. B. COE.

NEW YORK, August 6th, 1862.

Hon. S. C. BEMIS,—Dear Sir : Your favor of the 4th inst. is received, and I am profoundly grateful to you for the invitation it contains. The occasion to which you refer can hardly fail to be one of extraordinary interest, not only to the members of the Chapin stock, but to all who have the privilege to attend. I have been well acquainted with a large number of individuals in whose veins the Chapin

11

blood flowed, and I have never known one by whom the name was dishonored. It would afford me pleasure to see the faces of these and other representatives of this honored name, but it is probable that my official engagements will forbid.

I have forwarded your communication to Rev. Mr. Phoenix, and you may expect to hear from him soon.

Thanking you again for your courtesy, I remain with much respect, Yours truly,

D. B. COE.

LETTER FROM REV. ALEXANDER PHOENIX.

HARLEM, September 16th, 1862.

S. C. BEMIS, Esq,—Dear Sir: I promised myself the pleasure of writing you a letter, expressing my interest in the occasion of the family gathering, to take place in Springfield the present week; but my increasing bodily infirmities prevent me from sending you any thing, more than a mere line, at this time. I wish to express my own interest in the meeting, and the advantages which may arise from it, to the place where for eleven years I offered to those around me the blessings of the Gospel. To many of them bearing the name of Chapin, I became strongly attached, by many acts of kindness and love ; I have had the pleasure of meeting, in other parts of the world, many who were estimable, not only for the name, but for the character which I hope may ever attach to it wherever found. I hope that the meeting of this family may be a very general one, and that its results may long be remembered with gratitude. To all of Chicopee parish, with whom I have been acquainted, and with whom you may meet, will you present my most affectionate remembrance ; and accept for yourself and family the assurance of my sincere friendship.

Truly yours, in the bonds of Christian love,

ALEX. PHOENIX.

LETTER FROM JAMES LORIN CHAPIN.

LINCOLN, Mass., September 15, 1862.

To SECRETARY OF THE CONVENTION OF THE CHAPIN FAMILY, —Dear Sir: As a member of the Chapin Family I feel much interested in the proceedings of the Convention which is to be held this week in Springfield, and I had hoped to be able to attend it. But the condition of our country and distracted state of affairs will prevent my doing so, and at this late hour I attempt to communicate to you and through you to the Convention, such few items of intelligence as I think may be of interest to its members. I have in my possession the family Bible of my great-grandfather, who originated in what was then Longmeadow but now is a part of Monson, about one-half mile south or south-west of Palmer Depot.

This Bible contains his family record, most of it apparently in the handwriting of the owner, and I send you a copy of it for use if desired.* I trace my family to Luke the eldest of the family. He married and lived in Greenwich, Washington County, N. Y., and had two sons, Loring and James. When these children were quite young, he went to Canada with a drove of cattle, and so far as I can learn was never heard from afterwards. My father, when a lad of twelve years of age, came to an uncle's of his in Holland, Mass., and lived in Massachusetts until he married. He then went to New York State, and staid a few years. He then returned to Massachusetts, to the town of Sturbridge, Worcester County, where he continued to reside until about 1856, when he, with my youngest brother, went to Kansas, where they both died. He had a family of twelve children, of whom seven are known to be living, and one not known. If any

---

*COPY OF THE FAMILY RECORD OF LUKE CHAPIN.

Luke, born the twenty-third day of June, in the year 1762.
Justis, born the twenty-third day of February, in the year 1764.
Eli, born the twelfth day of February, in the year 1766, and died the fourteenth day of the same.
Oliver, born the seventeenth day of March, in the year 1767.
Elizabeth, born the twenty-ninth day of August, in the year 1769.
Ann, born the thirty-first day of March, in the year 1771.
Levi, born the thirty-first day of March, in the year 1773.
Mary, born the eighth day of May, in the year 1775.
Joseph, born the eighteenth day of September, in the year 1776.

The death of Mother Chapin, May 27, in the year 1770.
[In another handwriting.]
The death of my husband, September 30, in the year 1776.
The death of Joseph Chapin, January 24, in the year 1780.
The death of Oliver Chapin, December 9, 1791.

By a memorandum in another place it appears that Luke Chapin, father of the above family, died in the army, September 30, 1776.

information should be obtained at the Convention that will further trace the family connexion, I should be very happy to be notified of the fact. I am, very respectfully, yours,

JAMES LORIN CHAPIN.

LETTER FROM LYMAN CHAPIN.

The following letter, unaccountably delayed, from the late Lyman Chapin of Albany, possesses a melancholy interest, in the fact that, between the date on which it was written, and that on which it was received, the writer died and the great gathering of the family occurred. Mr. Chapin, while visiting a daughter in Missouri, wrote the letter, and afterwards returned, and, while on a visit to Long Branch, N. J., fell dead upon the beach, after a bath. The purity and integrity of Mr. Chapin's life give him a right to this recognition at the hands of the General Committee, of which he was a member:

JEFFERSON CITY, MISSOURI,
June 28, 1862.

S. C. BEMIS, ESQ.—Dear Sir : Yours of the 19th instant, reached me at this place, yesterday. It would afford me much pleasure to allow my name to be used in the "Chapin-Gathering" circular, agreeably to your request, and to "aid in bringing together the whole tribe, to their old New England home," were I to be so situated during the Summer, that such aid could be rendered, but it is now a matter of great uncertainty where I shall spend the next two or three months.

Hoping to be with you at the Gathering, I remain,

Yours, very respectfully,

LYMAN CHAPIN.

NAMES OF MEMBERS

CHAPIN FAMILY,

Who Attended the "Great Gathering" at Springfield, Mass., September 17th, 1862,

AS TAKEN FROM THE REGISTER.

———— ✦✦✦ ————

ABBOTT L. C., Eaton, Ohio.
Abbott Mrs., Eaton, Ohio.
Abbott ——, Eaton, Ohio.
Allen S., Chicopee.
Allen Mrs. Lois, Chicopee.
Allen D. A., Chicopee.
Allen Horace, Springfield.
Allen Mrs. H., Springfield.
Allen Lucy, Chicopee.
Arnold Submit Chapin, Somers, Ct.
Allen Mrs. Sophronia, Chicopee.
Allen Miss S. C., Chicopee.
Allen Joseph, Chicopee.
Allen Joshua, Springfield.
Allen Mrs. L. E., Springfield.
Allen Dexter C., Springfield.
Ainsworth O. W., Springfield.
Ainsworth Mrs. O. W, Springfield.
Ainsworth Carrie T., Springfield.
Ashley Roderick, Springfield.
Ashley Mrs. R., Springfield.
Allen Amzi, Mittineaque.
Allen Rhoda S., Mittineaque.
Allen Bennet, Haydenville.
Allen Cordelia, Haydenville.
Allen Mrs. Mather C., Springfield.
Allen D. A., Willimansett.
Ashley Mrs. A., West Springfield.
Allen E. O., Enfield, Ct.
Allen Mrs. E. O., Enfield, Ct.
Adams Mrs. Henry, Springfield.

BLISS GAD O., Longmeadow.
Bliss Mrs. Gad O., Longmeadow.
Burt Roderick, Springfield.

Burt Mrs. Roderick, Springfield.
Burt Sarah R., Springfield.
Burt Mary F., Springfield.
Burt Arthur S., Springfield.
Burt Moses, Springfield.
Bliss Elijah, Springfield.
Bliss Mrs. Elijah, Springfield.
Barbour Lucina, Hartford, Ct.
Barbour Harriet L., Hartford, Ct.
Bemis S. Augustus, Springfield.
Bemis Mrs. S. A., Springfield.
Bemis Arthur I., Springfield.
Bemis Mrs. Anna E., Springfield.
Bemis William C., Springfield.
Bemis Emma O., Springfield.
Bemis Stephen C., Springfield.
Bemis Mrs. Julia E., Springfield.
Bemis Miss Kate C., Springfield.
Bemis Thomas O., Springfield.
Bemis Henry S., Springfield.
Burnett Nelson W. South Hadley.
Burr Estes, Wilbraham.
Burr Mrs. Estes, Wilbraham.
Bugbee Jesse, South Hadley.
Bugbee Jesse, South Hadley.
Bugbee Horace, Springfield.
Bugbee Mrs. Horace, Springfield.
Bridges Luther, Warren.
Bridges Emeline, Warren.
Bridges James L., Warren.
Brewer M. E., Springfield.
Brewer T. A., Springfield.
Brewer Mrs. C. C., Springfield.
Brewer F. A., Springfield.
Black Orin, Swanzey, N. H.

Black Mrs. Orin, Swanzey, N. H.
Beckwith G. L., Providence, R. I.
Beckwith Mrs. G. L., Providence R. I.
Burt Wm., Longmeadow.
Bliss Mrs. Margaret C., Springfield.
Bailey S. E., Springfield.
Bailey Mrs. S. E., Springfield.
Bailey Miss Helen, Springfield.
Boylston Samuel, Springfield.
Boylston Mrs. Samuel, Springfield.
Bliss Luther, Springfield.
Bliss Mrs. Luther, Springfield.
Bliss Miss I.. Springfield.
Bliss James H., Springfield.
Bliss E., Springfield.
Bliss Harvey, Chicago, Ill.
Bliss Joel H., Springfield.
Bliss Mrs. W. A., Springfield.
Bliss Adelaide, Springfield.
Bliss Josephine, Springfield.
Bliss Julia, Springfield.
Bliss Everet, Springfield.
Ball Mrs. S. W., Springfield.
Bemis William L., Springfield.
Bemis Mrs. Mary C., Springfield.
Bemis Edward L., Springfield.
Butts Miss M. A., Springfield.
Burt Hannah, Springfield.
Brinsmade Wm. B. Springfield.
Brinsmade Mrs. C. B., Springfield.
Brinsmade John C., Springfield.
Brinsmade Anne L., Springfield.
Brinsmade William G., Springfield.
Brinsmade Samuel L., Springfield.
Brown Mrs. Amanda, Chester.
Bliss Margaret, Springfield.
Belding Ed. E., Chicopee.
Belding Mary L., Chicopee.
Bartlett Milo. Amherst.
Bliss Mary W., Longmeadow.
Bliss Mrs. Seth, Springfield.
Bliss Curtis, Springfield.
Bliss Mrs. Mary E., Springfield.
Bogardus Mrs. Betsy, Syracuse, N. Y.
Bogardus H. P., Syracuse, N. Y.
Bond Marvin, Greenwich.
Bond Mrs. Marvin, Greenwich.
Bliss Mrs. B. K., Springfield.
Bliss Samuel P., Springfield.
Bliss Elijah W., 2d, Springfield.
Bliss G. Theodore, Springfield.
Blair Samuel B., Warren.
Blair L. C., Warren.
Burbank A., Warren.
Broughton Miss L. B., Providence.
Buell S. K., Worcester.
Buell Mrs. S. K., Worcester.
Bartlett Asher, Chicopee Falls.
Bartlett Mrs. Asher, Chicopee Falls.

Bartlett Charles M., Chicopee Falls.
Barrett Charles H., Springfield.
Barrett Mrs. L. W., Springfield.
Breckenridge William L., Ware.
Breckenridge William C., Ware.
Ball E. H., Holyoke.
Ball Mrs. E. H., Holyoke.
Brigham Dexter C., Chicopee.
Brigham Mrs. D. C., Chicopee.
Brigham L. E., Chicopee.
Bonney O. E., South Hadley Falls.
Bonney Mrs. O. E., South Hadley Falls.
Bliss Austin, Monson.
Bliss Mrs. Austin, Monson.
Blodgett A. R., East Windsor, Ct.
Blodgett Mrs. A. R., East Windsor, Ct.
Bliss Luther, Jr., Springfield.
Bliss Hiram W., South Hadley Falls.
Bliss Mrs. H. W., South Hadley Falls.
Burns Mrs. John, Somers, Ct.
Birge Mrs. F. A., Agawam.
Burnett Enoch, Belchertown.
Bliss Mrs. E., Springfield.
Butterfield D. M., Chicopee.

Chapin A. C., Springfield.
Chapin Rose C., Springfield.
Chapin Adon A., Springfield.
Chapin Sidney N., Chicopee.
Chapin Ann, Chicopee.
Chapin Miss, Chicopee.
Chapin Mercy, Springfield.
Chapin D. E., Springfield.
Chapin Mercy H., Chicopee.
Chapin Julia, Chicopee.
Chapin Frederick L., Chicopee.
Chapin Luther, Pelham.
Chapin L. R., Pelham.
Chapin Eliza, Pelham.
Chapin Luther R., New Haven.
Chapin Sarah H., New Haven.
Chapin D. W., New Haven.
Chapin Emily, New Haven.
Chapin Lorilla A., New Haven.
Chapin Rachel A., New Haven.
Chapin Alanson, Agawam.
Chapin Electa E., Agawam.
Chapin Elthea, Agawam.
Chapin A. L., Springfield.
Chapin Susan M., Springfield.
Chapin Albert A., Springfield.
Chapin Marvin, Springfield.
Chapin Mrs. Marvin, Springfield.
Chapin Hattie S., Springfield.
Chapin Mary D., Springfield.
Chapin John M., Springfield.
Chapin Gratia R., Springfield.
Chapin E. S., Springfield.
Chapin Mrs. Louisa B., Springfield.

Chapin Henry W., Springfield.
Chapin Emma F., Springfield.
Chapin Alice S., Springfield.
Chapin Abijah W., Springfield.
Chapin Mrs. E. H., Springfield.
Chapin Frederick W., Springfield.
Chapin Edmund D., Springfield.
Chapin Addie L. G., Springfield.
Chapin Frank G., Springfield.
Chapin M. W., Warren.
Chapin Mrs. M. A., Warren.
Chapin Caleb F., Whitinsville.
Chapin Henry, Chicopee.
Chapin Experience, Chicopee.
Chapin Emily E., Chicopee.
Chapin Henry O., Chicopee.
Chapin Mrs. Lydia, Chicopee.
Chapin Thaddeus, Chicopee.
Chapin Mrs. Naomi, Chicopee.
Chapin Moses S., Gill, Mass.
Chapin Louis, Rochester, N. Y.
Chapin Mrs. Rachel S., Rochester, N. Y.
Chapin Ed. Dwight, Rochester, N. Y.
Chapin Louis S., Rochester, N. Y.
Chapin Wm. W., Rochester, N. Y.
Chapin Alice E., Rochester, N. Y.
Chapin Austin, Chicopee.
Chapin Mrs. Austin, Chicopee.
Chapin Henry W., Chicopee.
Chapin Maria A., Chicopee.
Chapin L. M., Chicopee.
Chapin Mrs. W., Chicopee.
Chapin Pliny, Chicopee.
Chapin Mrs. Pliny, Chicopee.
Chapin Sarah, Chicopee.
Chapin Norman, West Springfield.
Chapin Lydia, West Springfield.
Chapin Chalmers, West Springfield.
Chapin, Emelia B., West Springfield.
Chapin Edwin C., West Springfield.
Chapin Kate T., West Springfield.
Chapin Frank N., West Springfield.
Chapin Samuel D., Somers, Ct.
Chapin Sarah Phelps, Somers, Ct.
Chapin Horace D., Somers, Ct.
Chapin Mary D., Unionville, Ct.
Chapin Charles M., Unionville, Ct.
Chapin Sophronia, Waltham.
Chapin Augusta B., Lebanon, N. H.
Chapin Lucinda, Rochester, N. Y.
Chapin Josephine, Chicopee.
Chapin Mary, Chicopee.
Chapin Frederick, Chicopee.
Chapin William, Chicopee.
Chapin Eliphas I., Gill.
Chapin F. B., Camden, Me.
Chapin Japhet, Antwerp, N. Y.
Chapin Henry, Milford.
Chapin Mrs. J., Milford.

Chapin A. J., Springfield.
Chapin Anna C., Springfield.
Chapin George W., Springfield.
Chapin Mrs. M. G., Springfield.
Chapin Clarinda, Springfield.
Chapin Edgar I., Springfield.
Chapin Imogene A., Springfield.
Chapin Joseph C., West Springfield.
Chapin Mary M., West Springfield.
Chapin Seward B., West Springfield.
Chapin Evelina, West Springfield.
Chapin Margaret M., West Springfield.
Chapin Ashpel P., Ludlow.
Chapin Susan P., Ludlow.
Chapin Almi, Ludlow.
Chapin Andrew, Ludlow.
Chapin Julia, Ludlow.
Chapin Lester, Ludlow.
Chapin Hatty, Ludlow.
Chapin H. L., Springfield.
Chapin Lydia, Springfield.
Chapin Elias, Springfield.
Chapin M. W., Springfield.
Chapin Emily A., Springfield.
Chapin William H., Springfield.
Chapin C. L., Springfield.
Chapin Rev. A. L., Amsterdam, N. Y.
Chapin Doctor A., Winchester.
Chapin Miss E. D., Winchester.
Chapin Lewis, Worcester.
Chapin Mrs. Lewis, Worcester.
Chapin Edwin, Worcester.
Chapin Annetta M., Worcester.
Chapin Newman, Willimansett.
Chapin Mrs. Newman, Willimansett.
Chapin Ellen H., Willimansett.
Chapin Henry, Springfield.
Chapin Henry A., Springfield.
Chapin Mrs. H. A., Springfield.
Chapin Elizabeth, Springfield.
Chapin Mrs E., Springfield.
Chapin Miranda, Springfield.
Chapin Lovisa, Springfield.
Chapin Susan, Springfield.
Chapin E. A., Rutland, Vt.
Chapin Mrs. E. A., Rutland, Vt.
Chapin Hiram F., Granby, Ct.
Chapin Orphia, Granby, Ct.
Chapin Martha, Chicopee.
Chapin Adelaide, Chicopee.
Chapin John B., Chicopee.
Chapin Fanny L., Chicopee.
Chapin Olin L., Chicopee.
Chapin Sidney, Chicopee.
Chapin H. P., Chicopee.
Chapin Japhet, Chicopee.
Chapin T. L., Chicopee.
Chapin Thomas W., Chicopee.
Chapin Lucas E., Chicopee.

Chapin Angelina, Chicopee.
Chapin William D., Chicopee.
Chapin Mrs. Emily, Chicopee.
Chapin Ephraim, Waterloo, N. Y.
Chapin Mrs. Ephraim, Waterloo, N. Y.
Chapin Charles H., Chatham.
Chapin Theodore P., Tolland, Ct.
Chapin Amelia M., Tolland, Ct.
Chapin Mrs. William A., Springfield.
Chapin William A., Springfield.
Chapin William A. Jr., Springfield.
Chapin Martha H., Springfield.
Chapin Newman A., Springfield.
Chapin Sarah A., Springfield.
Chapin Miss E. W., Springfield.
Chapin William F., Springfield.
Chapin Miss O. S., Springfield.
Chapin William, Philadelphia.
Chapin John D., Canandaigua.
Chapin Moses, Rochester.
Chapin Mrs. Lucy T., Rochester.
Chapin Mr. S. B., Springfield.
Chapin Mrs. D. B., Springfield.
Chapin D. B., Springfield.
Chapin S. C., Springfield.
Chapin James O., Springfield.
Chapin George, Albany.
Chapin Sheldon, Seneca Falls.
Chapin Frances, New York.
Chapin Mrs. Bridgeman, Springfield.
Chapin Alpheus, Boston.
Chapin Mrs. Alpheus, Boston.
Chapin Lucas B., Chicopee.
Chapin Mrs. S., Chicopee.
Chapin A. D., Springfield.
Chapin Mrs. A. D., Springfield.
Chapin Josephine Chicopee.
Chapin Mrs. J., Chicopee.
Chapin Mrs. F. J., Chicopee.
Chapin M. J., Chicopee.
Chapin Augusta B., Westfield.
Chapin Mrs. S. C., Westfield.
Chapin George W., Westfield.
Chapin Giles S., Chicopee.
Chapin Mrs. G. S., Chicopee.
Chapin Julia A., Chicopee.
Chapin Wells M., Chicopee.
Chapin M. E., Chicopee.
Chapin Mrs. A. B., Hartford.
Chapin J. B., Hartford.
Chapin Albert, Hartford.
Chapin George, Hartford.
Chapin Orange, Willimansett.
Chapin Julia, Willimansett.
Chapin Daniel, East Windsor.
Chapin Mrs. A., East Windsor.
Chapin Harriet C., East Windsor.
Chapin Charles W., Springfield.
Chapin Mrs. Charles W., Springfield.

Chapin Charles K., Springfield.
Chapin Jeanette H., Springfield.
Chapin Warren, Holyoke.
Chapin Mrs. Warren, Holyoke.
Chapin Jane, Holyoke.
Chapin Asahel, New York.
Chapin H. J., Holyoke.
Chapin Eveline, Leyden.
Chapin Mrs. E., Leyden.
Chapin John M. Holyoke.
Chapin T. P., Holyoke.
Chapin Nathaniel, Westfield.
Chapin Levi, Walpole, N. H.
Chapin Heman, New Hartford.
Chapin Jonathan, Walpole.
Chapin Philip, Walpole.
Chapin Mary, Westfield.
Chapin C. M., New Hartford.
Chapin Nathaniel G., Boston.
Chapin Heman M., Cleveland.
Chapin Harriet L. Boston.
Chapin Cynthia M., Walpole.
Chapin E. M., New Hartford.
Chapin Mary E., New Hartford.
Chapin Mabel H., Boston.
Chapin Oliver R., East Longmeadow.
Chapin Mrs. O. R., East Longmeadow.
Chapin Titus, Chicopee.
Chapin Mrs. Titus, Chicopee.
Chapin Ed. E., Chicopee.
Chapin William D., Chicopee.
Chapin Mrs. William D., Chicopee.
Chapin Sumner, Chicopee.
Chapin Mrs. Sumner, Chicopee.
Chapin Abel, Chicopee.
Chapin Hon. Henry, Chicopee.
Chapin Rev. A. L., Beloit, Wis.
Chapin Henry L., Greenfield.
Chapin Mrs. Fanny A. Greenfield.
Chapin George C., Chicopee.
Chapin Mrs. Sophia A., Chicopee.
Chapin Neri, Chicopee.
Chapin Abby, Chicopee.
Chapin Oliver, Leyden.
Chapin Mrs. C. L., Leyden.
Chapin Lucina, Bernardston.
Chapin Mrs. H., Bernardston.
Chapin L. P., Bernardston.
Chapin Martha, Bernardston.
Chapin Harriet L., Bernardston.
Chapin Albert, Bernardston.
Chapin Caroline, Bernardston.
Chapin Louis J., Lee.
Chapin Mrs. L. J., Lee.
Chapin Louis F., Lee.
Chapin Morris G., Olena O.
Chapin Mrs. M. G., Olena O.
Chapin George F., Newport, N. H.
Chapin Samuel W., Bernardston.

Chapin Fred., Greenfield.
Chapin Mary, Greenfield.
Chapin William F., Springfield.
Chapin Miss T. F., Springfield.
Chapin William, Springfield.
Chapin Roxanna, Longmeadow.
Chapin Simeon R., Longmeadow,
Chapin Julia E., Longmeadow.
Chapin Nancy J., Gill.
Chapin George A., Boston.
Chapin Mrs. George A., Boston.
Chapin Miss Sarah D., Boston.
Chapin G. G., Boston.
Chapin Caleb, Greenfield.
Chapin John, Greenfield.
Chapin Smith, Worcester.
Chapin Warren, Staffordville.
Chapin Mrs. Warren, Staffordville.
Chapin W. E., Springfield.
Chapin Mrs. W. E., Springfield.
Chapin Ed. K., Springfield.
Chapin Rev. H. M., Ripon, Wis.
Chapin Mrs. H. M., Ripon, Wis.
Chapin Orlando, Springfield.
Chapin Alex., Springfield.
Chapin Mrs. E., Springfield.
Chapin Francis E., Springfield.
Chapin Margaret D., Springfield.
Chapin Phena E., Springfield.
Chapin Sarah W., Springfield.
Chapin Heman, Springfield.
Chapin Daniel M., Springfield.
Chapin Carrie B., Springfield.
Chapin Ziba, Cambridgeport, Vt.
Chapin Jacob, Westminster, Vt.
Chapin Henrietta, Albany.
Chapin L. C., New Haven.
Chapin Roxanna, Springfield.
Chapin D. D., Granby.
Chapin Ed., Granby.
Chapin Dennison, Granby.
Chapin Caroline A., Granby.
Chapin Mrs. D. D., Granby.
Chapin Miss E., West Springfield.
Chapin Harvey, South Hadley.
Chapin Mrs. Harvey, South Hadley.
Chapin Albert T., South Hadley.
Chapin Harriet, Chicopee.
Chapin Luther, Ashfield.
Chapin E. E., Ashfield.
Chapin Hervey, Holyoke.
Chapin Mrs. Hervey, Holyoke.
Chapin Persis P., Holyoke.
Chapin Amelia, Holyoke.
Chapin Sarah J., Holyoke.
Chapin J., Providence.
Chapin Mrs. J., Providence.
Chapin George W., Providence.
Chapin Mrs. George W., Providence.

Chapin Mrs. William G., Providence.
Chapin Ed., Providence.
Chapin Anna, Providence.
Chapin A. P., Granby.
Chapin Mrs. A. P., Granby.
Chapin Ed., Ludlow.
Chapin Mrs. Ed., Ludlow.
Chapin Ed. H., Ludlow.
Chapin Allison, Ludlow.
Chapin Hatty J., Ludlow.
Chapin J. B., Providence.
Chapin Charles B., Providence.
Chapin Mrs. J. B., Providence.
Chapin Miss L., Providence.
Chapin Miss J., Providence.
Chapin L. V. H., Ludlow.
Chapin Solomon, Malden.
Chapin Mary B., Ludlow.
Chapin F. L., Ware.
Chapin S. P., Springfield.
Chapin Mrs. S. P., Springfield.
Chapin Charles, Springfield.
Chapin Mrs. Charles, Springfield.
Chapin Emily C., Suffield.
Chapin J. B., Troy.
Chapin J. D., Troy.
Chapin Mrs J. B., Troy.
Chapin Miss J. E., Troy.
Chapin Norman, Ann Harbor.
Chapin Volney, Ann Harbor.
Chapin Mrs. Charles L., Warren.
Chapin Frank M., Warren.
Chapin Charles E., Chicopee.
Chapin William N., Chicopee.
Chapin Emma, Chicopee.
Chapin Miss Alice, Chicopee.
Chapin Caroline, Ludlow.
Chapin Julius E., Greenfield.
Chapin Mrs. J. E., Greenfield.
Chapin H. A., Greenfield.
Chapin Jason, Worcester.
Chapin Ed., New York.
Chapin Philo, Granby.
Chapin Mrs. Philo, Granby.
Chapin Helen A., Granby.
Chapin Charles O., Springfield.
Chapin Mrs. Charles O., Springfield.
Chapin Charles L., Springfield.
Chapin H. G., Springfield.
Chapin James L., New York.
Chapin Mrs. J. L., New York.
Chapin Miss C. A., New York.
Chapin Sarah, Springfield.
Chapin Mrs. E. S., Springfield.
Chapin Mary, Gill.
Chapin John A., Calais, Vt.
Chapin Mrs. John A., Calais, Vt.
Chapin P. H., Springfield.
Chapin Mrs. P. H., Springfield.

Chapin John A. Jr., Springfield.
Chapin R. S., New York.
Chapin Mrs. R. S., New York.
Chapin Julia, New York.
Chapin Alvah, Thompsonville.
Chapin J. E., Springfield, Ill.
Chapin M. W., Chicopee.
Chapin Mrs. M. W., Chicopee.
Chapin D. M., Ogdensburg.
Chapin Ed. E., Chicopee.
Chapin Mary, South Wilbraham.
Chapin Ambrosia, South Wilbraham.
Chapin Mrs. A., South Wilbraham.
Chapin Martin C., Holyoke.
Chapin Mrs. M. C., Holyoke.
Chapin Giles S. Jr., Granby.
Chapin Mrs. G. S. Jr., Granby.
Chapin E. W., Willimansett.
Chapin Mrs. W. Willimansett.
Chapin Marcus, Monson.
Chapin Enoch, South Hadley.
Chapin Mrs. H. J., South Hadley.
Chapin L. Dwight, Amsterdam.
Chapin Mrs. L. D., Amsterdam.
Chapin Lysander, Chicopee.
Chapin Mrs. L., Chicopee.
Chapin A., Chicopee.
Chapin Eleanor D., Chicopee.
Chapin Mrs. D. F., Chicopee.
Chapin Bryant, Chicopee.
Chapin William B., Chicopee.
Chapin G., Chicopee.
Chapin Mrs. G., Chicopee.
Chapin J. G., Somers, Ct.
Chapin Mrs. J. G., Somers, Ct.
Chapin Lysander, Holyoke.
Chapin Mrs. L., Holyoke.
Chapin H. L., Holyoke.
Chapin George M., Holyoke.
Chapin Mrs. William L., Springfield.
Chapin Mrs. Olive, Somers.
Chapin S. P., New York.
Chapin Mary W., South Hadley.
Chapin J. H. P., South Hadley.
Chapin Mrs. J. H. P., South Hadley.
Chapin Samuel W., Chicopee.
Chapin Mrs. Samuel W., Chicopee.
Chapin Emma R., Chicopee.
Chapin H. E., Chicopee.
Chapin Mrs. C. M., Enfield, Ct.
Chapin Mrs. H., Enfield, Ct.
Chapin Edwin, Hadley.
Chapin A., New York.
Chapin Elijah, Palmer.
Chapin Lucinda, Palmer.
Chapin S. C. S., Chicopee.
Chapin Daniel E., Boston.
Chapin Silas W., South Hadley.
Chapin Ralph S., Wilbraham.

Chapin Mrs. R. S., Wilbraham.
Chapin Samuel, Wilbraham.
Chapin Ellen E., Wilbraham.
Chapin C. W., Springfield.
Chapin Mrs. C. W., Springfield.
Chapin Daniel T., Enfield.
Chapin Francis, Enfield.
Chapin H. R., Enfield.
Chapin J. T., Enfield.
Chapin Mrs. T., Enfield.
Chapin Eliza, Enfield.
Chapin Mrs. D. M., Enfield.
Chapin Mrs. H. R., Enfield.
Chapin E. P., Springfield.
Chapin Milo, Springfield.
Chapin Abby, Springfield.
Chapin Emma, Springfield.
Chapin Julia A., Springfield.
Chapin William W., Providence.
Chapin Elias F., Belchertown.
Chapin Mrs. E. F., Belchertown.
Chapin Ed. M., Belchertown.
Chapin Pliny, Springfield.
Chapin Mrs. Pliny, Springfield.
Chapin E. P., Springfield.
Chapin Elam, Hartford.
Chapin Andrew P., Chicopee Falls.
Chapin Mrs. A. P., Chicopee Falls.
Chapin Harriet, Chicopee Falls.
Chapin Fred. Chicopee Falls.
Chapin Charles A., Hartford.
Chapin Hollis T., Hartford.
Chapin Mrs. Hollis T., Hartford.
Chapin V. H., East Longmeadow.
Chapin Mrs. V. H., East Longmeadow.
Chapin T. B., Enfield, Ct.
Chapin Mrs. T. B., Enfield Ct.
Chapin Harriet E., Enfield, Ct.
Chapin Francis S., Pittsfield.
Chapin Phineas, Great Barrington.
Chapin R. V., Hazardville.
Chapin Miss S. O., Longmeadow.
Chapin D. C., Henrietta, N. Y.
Chapin Charles S., Worcester.
Chapin M. W., Hartford.
Chapin John R., Rahway, N. J.
Chapin Miss M. A., Chicopee.
Chapin Mrs. S., Hartford.
Chapin Hannah E., Hartford.
Chapin Novadus N., E. Longmeadow.
Chapin J. F., Hartford.
Chapin Ellen M., Hartford.
Chapin Charles P., Boston.
Chapin N. H., Charlestown.
Chapin George F., Charlestown.
Chapin Mrs. M., Rochester.
Chapin Dr. E. R., Flatbush, L. I.
Clark Mrs. N. B., Springfield.
Clark N. B., Springfield.

Call Amos, Springfield.
Call Mrs. Amos, Springfield.
Call Charles A., Springfield.
Call George N., Springfield.
Call Ruema C., Springfield.
Calhoun William B., Springfield.
Calhoun Mrs. Margaret, Springfield.
Calhoun Martha C., Springfield.
Calhoun Charles K., Springfield.
Childs A. H., Holyoke.
Childs Mrs. P. S., Holyoke.
Childs M. L., Chicopee.
Capron Samuel M., Hartford.
Capron Mrs. Samuel M., Hartford.
Cutler Mrs. Francis, Akron, O.
Clark Julia A., Springfield.
Crane Lois A., Springfield.
Cornwall Royal, Springfield.
Cornwall Mrs. Olena, Springfield.
Chandler Mrs. L. W., Cazenovia, N. Y.
Chandler A. C., Cazenovia, N. Y.
Chilson Otis, Enfield, Ct.
Chilson Mrs. Otis, Enfield, Ct.
Chilson Myra, Enfield, Ct.
Chapman G. H., Chicopee.
Conkey Katie, Rochester.
Colton Emma, Springfield.
Colton Harriet, Huntington.
Collins E. W., Westfield.
Colton Horatio, Chicopee.
Colton Mrs. H., Chicopee.
Colton Julia, Chicopee.
Colton Mary, Chicopee.
Clark Mrs. William H., Springfield.
Clark Mrs., Springfield.
Cooley Mrs. L. E., West Springfield.
Collins Mrs. L. M., Hartford.
Clough Mrs. P. F., Chicopee.
Collins S. H., Worcester.
Colton A. M., Easthampton.
Colton Mrs. A. M., Easthampton.
Colton Lucy M., Greenwich.
Colton Sarah, Greenwich.
Colton Augusta, Greenwich.
Collins J., Somers.
Collins Mrs. J., Somers.
Collins Noah C., Somers.
Collins Mrs. N. C., Somers.
Collins D. F., Somers.
Collins Martha C., Somers.
Cooley Mrs. A., Longmeadow.
Cooley A., Longmeadow.
Cooley Martha, Longmeadow.
Clapp Martha, Springfield.
Clapp Mrs. N., Springfield.
Clapp E. H., Springfield.
Clapp Mary C., Springfield.
Clapp A. M., Springfield.
Clapp D. M., Westhampton.

Clapp Mrs. L. F., Westhampton.
Chaffee Mrs. D. D., South Wilbraham.
Chaffee Lucy M., South Wilbraham.
Chaffee D. D., South Wilbraham.
Collins Wm. O., Somers.
Cowles Harriet, Westfield.
Cowles Newell, Westfield.
Cowles R. C., Westfield.
Colton Mrs. Ed. K., Longmeadow.
Chilson Mrs. C., Enfield.
Crooks J. W., Springfield.
Crooks Mrs. J. W., Springfield.

Day Albert, Hartford.
Day Harriet Chapin, Hartford.
Day Albert S., Hartford.
Day Caroline B., Hartford.
Douglass Henry, Norwich, Ct.
Douglass Mrs. E. W., Norwich, Ct.
Davis Harriet A., Hartford.
Dewey Mrs. H., Springfield.
Dewey Miss Helen, Springfield.
Davis Samuel, Prescott,
Davis Mrs. Samuel, Prescott.
Davis Orland, Stafford.
Davis William O., Stafford.
Dillaber William, Easthampton.
Dillaber Jesse, Chicopee.
Dillaber William J., Chicopee.
Davis D., West Stafford.
Davis George, Stafford.
Davis Franklin C., Stafford.
Darling M. E., Greenfield.
Darling Mrs. M. E., Greenfield.
Darling Mary, Greenfield.
Davis Miss M. S., West Stafford.
Davis Mrs. D. E., West Stafford.
Davis Noah C., West Stafford.
Davis Daniel, Somers.
Davis Ellen S., Greenwich.
Davis Lucy S., Greenwich.
Davis O., Stafford Springs.
Davis Mrs. O., Stafford Springs.
Davis Mrs. L., Stafford Springs.
Davis Mary, Stafford Springs.
Davis E., Stafford Springs
Danks R., Chicopee.
Danks Mrs. R., Chicopee.
Davis Lorenzo, Greenwich.
Danks S. E., Chicopee.
Day Newton, Willimansett.
Davis Laura M., Springfield.
Davis Spencer, Somers.
Davis Mary W., Somers.
Davis Sarah W., Somers.
Davis Laura A., Somers.
Day Harriet, Newark, N. J.

Edwards Mary, Springfield.

Ely Nancy, Westfield.
Ely Joseph, Holyoke.
Ely Mrs. Joseph, Holyoke.
Ely Austin, Holyoke.
Ely Mrs. Austin, Holyoke.
Ely M. A., Holyoke.
Ely Emma, Holyoke.
Ely Rev. J., Thompsonville.
Ely Mrs. J., Thompsonville.
Ely Samuel, Holyoke.
Ellis W. W., Stafford.
Ellis Mrs. W. W., Stafford.
Ely Cotton, West Springfield.
Ethrington James, Hartford.
Ethrington Mrs. James, Hartford.

FIRMAN FRANCIS B., South Wilbraham.
Fairbanks Ashel, Warren.
Fairbanks, Mrs. M. C., Warren.
Fairbanks William H., Warren.
Ferry Louis, Easthampton.
Ferry Eli, Chicopee.
Ferry Miss A. C., Chicopee.
Foot Mrs. Hannah, Springfield.
Ferry Henry A., Springfield.
Ferry Mrs. H. A., Springfield.
Fisher Emily C., Antwerp, N. Y.
Forbes George F., West Springfield.
Forbes Mrs. George F. W., Springfield.
Fish L. B., South Hadley.
Fish Mrs. L. B. South Hadley.
Fiske Mrs. R. C., Upton.
Flagg Edward, Springfield.
Flagg Mrs. E., Springfield.
Flagg E. Jr., Springfield.
Fuller Joseph, Suffield.
Fuller H. S., Suffield.
Ferry E., Easthampton.
Foote Hon. Solomon, Rutland, Vt.
Frost Elizabeth M., East Windsor.
Frost Mrs. H. S., East Windsor.
Frost Ella A., East Windsor.
Ferry C. B., Becket.
Ferry Mrs. C. B., Becket.
Foster Mrs. B. C., Springfield.

GRAVES MARY, Hatfield.
Gardner Lois, Springfield.
Gardner Emma, Springfield.
Gardner Abby, Springfield.
Goodman Sarah J., Springfield.
Gaylord Moses, South Hadley.
Gaylord Mrs. Moses, South Hadley.
Guyle Mrs. G., Springfield.
Guyle Albert, Springfield.
Green Daniel, South Coventry.
Green Emma, South Coventry.
Green Mrs. Daniel, South Coventry.
Goodman Mrs. P., Springfield.

Graves Levi, Springfield.
Graves Mrs. L., Springfield.
Graves Maria, Springfield.
Graves Myron, Springfield.
Gleason Abner C., West Springfield.
Gleason Mrs. A. C., West Springfield.
Green F. C., Northampton.
Green Mrs. F. C., Northampton.
Griswold F. J., Chicopee.

HARDING J. W., Longmeadow.
Harding Mrs. J. W., Longmeadow.
Horton Samuel, Westfield.
Horton Mrs. Samuel, Westfield.
Horton Clara, Westfield.
Horton Sophia, Westfield.
Hamilton Mrs. L., Chicopee.
Hinckley Rufus, South Hadley.
Hinckley Mrs. Rufus, South Hadley.
Hinckley Hattie S., South Hadley.
Hitchcock L., Springfield.
Hitchcock Mrs. Mary, Springfield.
Hitchcock Roswell, Shelburne Falls.
Hitchcock Walter, Wilbraham.
Hitchcock Lucy R., Wilbraham.
Hitchcock H. Louisa, Wilbraham.
Hitchcock Charles E., Wilbraham.
Hitchcock Duane W., Wilbraham.
Hitchcock Fred. R., Wilbraham.
Henry Lucy A., Chicopee.
Hough Harriet B., Westfield.
Hancock Moses, Springfield.
Hancock Mrs. P., Springfield.
Hancock Miss M., Springfield.
Hitchcock Stephen Mrs., Springfield.
Hitchcock Clara M., Springfield.
Hill P., Boston.
Hill Mrs. Ellen, Boston.
Hodge Mrs. S., West Hartford.
Hamilton O. M., Chicopee.
Harris H. H., Chicopee.
Harris Mrs. H. H., Chicopee.
Harris F. H., Springfield.
Harris Mrs. F. H., Springfield.
Harris Mary C., Springfield.
Harris Fred. Springfield.
Hunt Mrs. H. P., Boston.
Hazen Elbridge, Springfield,
Hazen Mrs. Elbridge, Springfield.
Harvey Rhoda A., Chesterfield.
Hoyt R. H., Bernardston.
Hoyt Adeline H., Bernardston.
Hatfield Mrs. A., Springfield.
Hatfield Lucy A., Springfield.
Hitchcock Lucina, Springfield.
Hitchcock Mrs. L., Springfield.
Hitchcock William O., Springfield.
Holland J. G., Springfield.
Holland Mrs. J. G., Springfield.

Holland Annie E., Springfield.
Holland Kate M., Springfield.
Holland Theodore, Springfield.
Hodgekin Charles, Prescott.
Hodgekin Mrs. Charles, Prescott.
Hancock James, Springfield.
Hunt Mrs. J. S., Boston.
Hart Mrs., Springfield.
Hitchcock W. L., Chicopee.
Hitchcock Mrs. W. L., Chicopee.
Hitchcock Mary L., Chicopee.
Hitchcock Alden, Springfield.
Hitchcock R. D., New York.
Hitchcock Henry, Springfield.
Hitchcock Sophia, Springfield.
Hitchcock D. B., Holyoke.
Hitchcock Mrs. D. B., Holyoke.
Henderson James, Enfield.
Henderson Mrs. James, Enfield.
Herrick Mary C., Westfield.
Hinchman F. H., Detroit.
Hinchman John M., Detroit.
Herrick Ed. M., Westfield.
Herrick Mrs. Mary, Westfield.
Haile Henry W., Hinsdale, N. H.
Haile Mrs. H. W., Hinsdale, N. H.

Isham John M., South Wilbraham.
Isham Mary E., South Wilbraham.
Isham Fanny A., South Wilbraham.

Jameson A., Chicopee.
Jameson Mrs. A., Chicopee.
Judd Harvey, South Hadley.
Judd Mrs. Harvey, South Hadley.
Jocelyn Sarah, Springfield.
Judd W. S., South Hadley.
Judd J. D., South Hadley.
Judson Willard, Uxbridge.
Jones S., Ludlow.
Jones Mrs. J., Ludlow.
Jones Daniel, Ludlow.
Jones Mrs. D., Ludlow.
Jones H. S., Ludlow.
Jones Mrs. H. S., Ludlow.
Jones Miss S., Ludlow.
Jones Miss P., Ludlow.
Jones Miss J., Ludlow.
Judd H. H., South Hadley Falls.
Judd Mrs. H. H., South Hadley Falls.

Kingsbury M., Arkadelphia, Akr.
Kingsbury Abel C., Hartford.
Kellogg J. E., South Hadley.
Kellogg Mary W., South Hadley.
Kellogg Hattie S., South Hadley.
Kellogg Amos, South Hadley.
Kellogg Cynthia, Providence.
Kellogg Mrs. E. C., Chicopee.

Kellogg M. S., Chicopee.
Kingler Mrs. S. J., Cleveland, Ohio.
Kingler William O., Cleveland, Ohio.
King William B., Suffield.
King Mrs. William B., Suffield.
King S. Z., Warehouse Point.

Lenoir William, Springfield.
Lenoir Mrs. William, Springfield.
Lathrop Benjamin, Springfield.
Lathrop Mrs. Benjamin, Springfield.
Lee Mrs. William, Springfield.
Lee Roswell, Springfield.
Lee Samuel K., Springfield.
Lombard J. B., Warren.
Lombard Mrs. A. A. C., Warren.
Lombard Mary C., Warren.
Lombard Edward, Warren.
Lombard Elsie, Springfield.
Lombard Delia, Springfield.
Lombard Zelotes, Springfield.
Lombard Sarah Jane, Springfield.
Lombard Edward M., Springfield.
Lord Mrs. A. G., Springfield.
Loring Henry M., Chicopee.
Loring Mrs. H. M., Chicopee.
Lamb Melissa, South Hadley.
Lamb A. G., South Hadley.
Loring Betsy, Chicopee.
Lathrop Mrs. P., South Hadley.
Lamb George E., South Hadley.
Lamb Mary, South Hadley.

Mather Timothy, Hartford.
Mather Nancy, Hartford.
Medlicott W. G., Longmeadow.
Medlicott Mrs. E. G., Longmeadow.
Morris Judge O. B., Springfield.
Morris Lydia, Springfield.
Morris Maria M., Springfield.
Morris Robert O., Springfield.
Morris Henry, Springfield.
Morris Mrs. Henry, Springfield.
Morris Mary W., Springfield.
Morris Edward, Springfield.
Morris Fred W., Springfield.
Morris George B., Springfield.
Morris Mrs. George B., Springfield.
Morris Carrie, Springfield.
Moody John, Ludlow.
Moody Mrs. S. C., Ludlow.
Moody Hannah, Ludlow.
Mills Samuel P., Springfield.
Mills Ellen, Springfield.
Morris R. D., Springfield.
Morris Hattie B., Springfield.
Morris C. S., Springfield.
Monson Mary S., Cazenovia, N. Y.
Mattack Mrs. E. G., Alton, Ill.

Mattack John C. Alton, Ill.
Mattack Jennie, Alton, Ill.
Montague Mary, Granby.
Monson G., Huntington.
McElwain D., North Becket.
McElwain Mrs. D., North Becket.
McElwain Charles, North Becket.
Montague Elliot, South Hadley.
Montague Mrs. E., South Hadley.
Monson Mrs. G., Huntington.
Monson M. A., Huntington.
Monson Emma H., Huntington.
Monson Clara F., Huntington.
Marshall·Orange, Westfield.
Metcalf Mrs. Eben, Chardin, O.
Metcalf Miss. F., Thompsonville.
Merrick F. T., West Springfield.
Merrick Julia C., West Springfield.
Merrick E. J., West Springfield.
Moodey M. K., Brooklyn.
Moodey Hannah M., Brooklyn.
May Mrs. George, Cleveland.
May Minnie, Cleveland.
Moody L. A., Springfield.
Moody Mrs. L. A., Springfield.
Moody M., Springfield.
Moody Kate R., Springfield.
Moody Anna E., Springfield.
Moore J. H., Warren.
Moore Mrs. J. H., Warren.
Monson A., Monson.
Monson Mrs. A., Monson.
Monson Lucy P., Monson.
Morgan J. W., Greenwich.
Morgan Mrs. J. W., Greenwich.
Moseley H. E., Springfield.
Moseley Mrs. S. R., Springfield.
Moseley C. L., Springfield.
Moseley Nellie B., Springfield.
McKinstry A. L., Chicopee.
McKinstry Miss E., Chicopee.
Moffat Daniel A., Agawam.
Moffat Mrs. Daniel A., Agawam.
Moffat E. L., Agawam.
Marshall Francis, Westfield.
Marshall Elizabeth, Westfield.
Moffat Charles O., Agawam.
Moffat C. A., Agawam.
McKinstrey Rev. J. A., Harrington.
McKinstrey, Mrs. J. A., Harrington.
Morris J. F., Hartford.
Morris John B., South Wilbraham.
Morris Wm. P., South Wilbraham.
Morris Caroline, South Wilbraham.
Morris Elizabeth L., South Wilbraham.
Morton Mrs. L., Worcester.
Miller Joel, South Hadley Falls.
Miller Tryphena, South Hadley Falls.

NUTTING Mrs. M., Greenfield.
Nutting Mary E., Greenfield.
Norris Henry, Springfield.
Norris Mrs. P. C., Springfield.
Norris Ellen J., Springfield.
Norris Emma E., Springfield.
Nichols Mary E., Brimfield.
Newell J. C. B., Springfield.
Newell Lucy B., Springfield.
Nash Mrs. Simeon., South Hadley.
Norton D. W., Suffield.
Norton Mrs. D. W., Suffield.
Nicholson Harriet, Chicopee.

OLMSTEAD ISAAC, New York.
Olmstead Mrs. Francis, New York.

POMEROY OWEN, Somers, Ct.
Pomeroy Mrs. Owen, Somers, Ct.
Pomeroy Miss L., Somers, Ct.
Pomeroy Miss S., Somers, Ct.
Payne Mary Jane, Unionville, Ct.
Parker Lucinda, Springfield.
Preston Eliza, New York.
Pratt Edmund, Longmeadow.
Pratt Mrs. Sarah, Longmeadow.
Pratt Sarah, Longmeadow.
Pratt Mary, Longmeadow.
Pease Walter, Springfield.
Pease S. C., Springfield.
Pease L. W., Springfield.
Pease Jane C., Springfield.
Potter Jennett, Greenfield.
Patch Ely H., Springfield.
Patch Lucy A., Springfield.
Patch Geo. B., Springfield.
Patch Mary D., Springfield.
Patch E. B., Springfield.
Pendleton Miss H. D., Willimansett.
Parker A. G., Chicopee.
Parker Mrs. A. G., Chicopee.
Parker J. A., Chicopee.
Pease C. H., Chicopee.
Pease Mrs. C. H., Chicopee.
Pease Miss E., Chicopee.
Pease B. F., Nevada City.
Pease Marshall, Chicopee.
Pease Samuel C., Chicopee.
Pease M. C., Chicopee.
Pearson Mrs. Dr. Chicopee.
Pease Chas. N., Chicopee.
Pease Mrs. C. N., Chicopee.
Pease Clifford B., Chicopee.
Park Rodney, Bernardston.
Park Mrs. R., Bernardston.
Pease Theo. W., Thompsonville.
Pike Orrin, Springfield.
Pease Levi. S., Thompsonville.
Pease Susan, Thompsonville.

Parsons J. D., Hartford.
Parsons Mrs. J. D., Hartford.
Puffer Wm., Monson.
Puffer Mrs. Wm., Monson.
Pratt Abby C., Worcester.
Paige E. F., Chicopee.
Paige Mrs. E. F., Chicopee.
Packard Mrs. F. C., Monson.
Pease M. S., Chicopee.
Parmalee O. R., South Hadley.
Parmalee Mrs. O. R., South Hadley.
Pease Mary G., South Wilbraham.
Pease Mrs. E., Somers, Ct.
Phelps Roswell, Wilbraham.
Phelps Mrs. Roswell, Wilbraham.
Phelps Miss S. Wilbraham.
Phileo Mrs., Suffield.
Pendleton John, Willimansett.
Pendleton Susan S., Willimansett.
Parmalee Moses P., Underhill, Vt.
Puffer Mrs. Wm. H., Monson.
Pendleton Rodney, Chicopee.
Pendleton Mrs. Rodney, Chicopee.
Pryor Mary, Windsor.
Pease H., Springfield.

Remington Mrs. R., Springfield.
Rice Mrs. Lucy, Springfield.
Rumrill J. S., Chicopee.
Rumrill, Mrs. J. S., Chicopee.
Robinson J. C. Springfield.
Robinson Flavia J., Springfield.
Robinson H. A., Springfield.
Robinson Margaret L., Springfield.
Robinson Frank H., Springfield.
Richardson S. B., Warren.
Richardson Mrs. W. B., Warren.
Robinson J. S., Springfield.
Robinson Amelia A., Springfield.
Robinson E. C., Springfield.
Robinson Sarah, Springfield.
Robinson H. S., Springfield.
Robinson M. P., Springfield.
Robbins Mrs. R., Springfield.
Rumsey M. H., Springfield.
Rumsey Mrs. M. H., Springfield.
Rice Chas. W., Springfield.
Ray Helen M., Westfield.
Ray Edward A., Westfield.
Rice D. B., Springfield.
Rice Mrs. D. B., Springfield.
Rice Miss L. M., Springfield.
Russell Jeanett B., South Hadley.
Rice Mrs. Nathan, Wilbraham.
Rice N. C., Wilbraham.
Robinson Emeline, Springfield.
Robinson Fred. M., Springfield.
Rice Mrs. Daniel, Belchertown.
Reed Dr. S., Pittsfield.

Reed Mrs. S., Pittsfield.
Rockwell Dr. Owen, Westfield.
Rockwell Jarvis, Hinsdale.
Reynolds Miss H., Springfield.
Rumrill Mrs. R., Chicopee Falls.

Smith Wm. L., Springfield.
Smith Mrs. Wm. L., Springfield.
Sheldon Chas. C., Somers, Ct.
Stebbins Seth, Chicopee.
Stearns Sophia E., Lebanon.
Stebbins Melissa, Chicopee.
Stebbins Joseph O., Chicopee.
Stebbins Cynthia, Chicopee.
Skeele Otis, Willimansett.
Skeele Mrs. Otis, Willimansett.
Skeele Adeline M., Willimansett.
Skeele John O., Willimansett.
Sturtevant Warner C., Springfield.
Sturtevant Mrs. Julia E., Springfield.
Stearns Josiah, Chicopee.
Stearns Mrs. Mary, Chicopee.
Stearns Mary, Chicopee.
Smith Wm., South Hadley.
Smith Mrs. Wm., South Hadley.
Switzer Elizabeth, Warren.
Sykes Henry M., Suffield.
Sykes Mrs. H. A., Suffield.
Sykes Miss P. M., Suffield.
Southworth C. A., Ludlow.
Southworth Mrs. S. M., Ludlow.
Sanderson Harvey, Springfield.
Sanderson Edward, Springfield.
Smith Sarah, West Springfield.
Smith Rhoda C., West Springfield.
Smith John D., West Springfield.
Smith Susan B., West Springfield.
Smith Daniel, West Springfield.
Smith Mrs. Daniel, West Springfield.
Smith Miss S. E., West Springfield.
Smith Edward C., West Springfield.
Smith Geo. E., West Springfield.
Smith Henry A., West Springfield.
Sawyer Mrs. A. D., Springfield.
Smith S. C., Westfield.
Smith Mrs. S. C., Westfield.
Smith Miss Ellen S., Westfield.
Safford R. T., Springfield.
Safford Mrs. R. T., Springfield.
Safford Fanny, Springfield.
Safford Clara, Springfield.
Stebbins Mrs. Calvin, Springfield.
Stebbins Eliza B., Springfield.
Stedman P., Chicopee.
Stedman Mrs. P., Chicopee.
Stedman Edward P., Chicopee.
Spring Mrs. Josiah, Whitinsville.
Spooner J. C., Springfield.
Spooner, Mrs. P. T., Springfield.

Spooner J. M., Springfield.
Spooner Mrs. J. C., Springfield.
Spooner Frank E.. Springfield.
Stillman Levi S., Wethersfield.
Stillman Mrs. Levi, Wethersfield.
Smith Herman, South Hadley.
Smith Mary, South Hadley.
Smith Emily W., South Hadley.
Stacey Wm., South Hadley.
Stacey Miss M. J., South Hadley.
Storey Wm. H., New York.
Sheldon S. A., West Stafford.
Stevenson Ed. C., North Wilbraham.
Stevenson Mrs. E. C., No. Wilbraham.
Stockwell Anna P., Upton.
Stearns J. Jr., Northampton.
Stearns Mrs. J., Northampton.
Skeele Henry E., Springfield.
Skeele Mrs. H. E., Springfield.
Stearns Miss C., Northampton.
Stearns Adelaide, Northampton.
Smith Austin, Granby.
Smith Mrs. Austin, Granby.
Spooner Mrs. H. B., Worcester.
Smith S. F., Westfield.
Smith F. F., Westfield.
Smith Mary C., Westfield.
Strickland Mrs. P. C., Palmer.
Sanford Mason, Belchertown.
Smith Miss A., Amherst.
Smith Ed., South Hadley.
Smith J. M., Holyoke.
Smith Mrs. J. M., Holyoke.
Sanford George, Belchertown.
Sanford Sophia, Belchertown.
Sanford Nancy, Belchertown.
Stevens Ed., Chicopee.
Stevens Mrs. E., Chicopee.
Stevens Miss H., Chicopee.
Stevens Geo. E., Chicopee.
Stevens Chas. N., Chicopee.
Stebbins Miss S., Wilbraham.
Stebbins Miss H., Wilbraham.
Smith Sam'l C., Westfield.
Stedman B. H., Chicopee.
Stedman Mrs. B. H., Chicopee.
Stedman Mrs. S., Chicopee.
Smith D. S., Chicopee.
Storrs Mrs., Wilbraham.
Stebbins J. B., Springfield.
Stebbins Mrs. J. B., Springfield.
Storrs E. W., Longmeadow.
Storrs R. S., 3d, Longmeadow.
Stebbins L. W., Springfield.
Smith Quartus, Chicopee.
Stanley Mrs. A., West Hartford.
Simons Mary, West Hartford.
Scripture Emily, Wilbraham.
Scripture Abby, Wilbraham.

Smith Mrs. Dell, Easthampton.
Smith Lawrence, Easthampton.

THAYER GEO. W., Springfield.
Thayer Mrs. Geo. W., Springfield.
Thayer Emma, Springfield.
Thayer George, Springfield.
Thayer A. J., Charlestown.
Taylor E. S., Springfield.
Taylor Mrs. E. S., Springfield.
Taylor Sylvester C., Chicopee.
Taylor Sarah, Chicopee.
Taylor A. C., Chicopee.
Taylor Louisa B., Chicopee.
Taylor Geo. S.. Chicopee.
Taylor A. B., Chicopee.
Taylor Varnum N., Chicopee.
Taylor Elizabeth C., Chicopee.
Taylor Chas. A., Chicopee.
Taylor Jane D., Chicopee.
Taylor James E., Chicopee.
Taylor Electa B., Chicopee.
Taylor Wm. O., Boston.
Taylor Mary B., Boston.
Town Orange C., Willimansett.
Town Mrs. E. S., Willimansett.
Temple Eliza, Chicopee Falls.
Temple R., Chicopee Falls.
Town Wm., Fair Haven.
Town Milton, Chicopee.
Thompson Wm. H., East Windsor.
Thompson Mrs. Wm. H., E. Windsor.
Thornton Ruth A., Worcester.

VINTON J. B., Springfield.
Vinton Mrs. J. B., Springfield.
Vinton E. B., Springfield.
Vinton C. E., Springfield.
Vaile Dr. Joseph, Springfield.

WHITNEY Mrs. J. D., Fitchburg.
Whitney Miss L. J., Fitchburg.
Warner Mrs. B. F., Springfield.
White Myra, Springfield.
Waite Russell, Hatfield.
Waite Mrs. Russell, Hatfield.
Waite Chauncey S.. Chicopee.
Waite Mrs. C. S., Chicopee.
Waite Albert A., Chicopee.
Waite Mrs. A. A., Chicopee.
Waite F. R., Chicopee.
Waite Mary, Chicopee.
Woodward Marcus, Somers.
Woodward Newton D., Somers.
Woodward Mary M., Somers.
Webster Chas. P., Hartford.
Webster Margaret, Hartford.
Weeks James L., Warren.
Weeks Mattie E., Warren.

Whiting Mrs. Paul, Whitinsville.
Whiting Ann L., Whitinsville.
Ward Mrs. N., Chicopee.
Warren J. C., Eggertsville, N. Y.
Wellman Warren, Springfield.
Wellman Mrs. M., Springfield.
Wellman Abigail, Springfield.
Wellman Emma, Springfield.
White Mrs. R. C., Syracuse.
Wells Jerome, Chicopee.
Wells Mrs. L. C., Chicopee.
Wells Geo. L., Chicopee.
Webster Jed. P., Enfield.
Webster Mrs. J. P., Enfield.
West Bailey, Chicopee.
West Ann S., Chicopee.
Winans Dr. S., Springfield.
Winans Mrs. S., Springfield.
Winans Dormer C., Springfield.
Winans A. O., Springfield.
Wright Eber, Chicopee.
Wright Mrs. Eber, Chicopee.
Ward Mrs. Josiah, Amherst.
Ward Wm. F., Boston.
Warner Chas. P. L., Springfield.
Warner Mrs. C. P. L., Springfield.
Winans Mrs. Niles A., Springfield.
Wadsworth John A., Springfield.
Wadsworth Mrs. J. A., Springfield.
Ware Mrs. A. P., Springfield.
Waller Mary, Philadelphia.
Wright Mrs. S., Agawam.
Wright Geo. F., Agawam.
Wright Josephine A., Agawam.
Wright Wm. H., Springfield.

Wright Mrs. Wm. H., Springfield.
Wirke Mrs. H., Boston.
Wood Sarah C., Worcester.
Wood Mrs. E. C., Upton.
Wallace Mrs., Philadelphia.
Wood Chas., Agawam.
Wood Mrs. Chas., Agawam.
Wright H. J., New York.
Warriner A., West Springfield.
Warriner Mrs. A., West Springfield.
Warriner Francis, Chester.
Worthington L. N., Agawam.
Worthington Martha C., Agawam.
Worthington Mary L., Agawam.
Wolcott Helen O., Holyoke.
Wolcott James M., Holyoke.
Wolcott Mrs. J. M., Holyoke.
Whitin Henry, Boston.
Wood Gaius, Somers.
Wood Mrs. G., Somers.
Wright J. C., Holyoke.
Wright Mrs. J. C., Holyoke.
Wood H. S., Greenfield.
Warren J. B., Wilbraham.
Warren Betsey, Wilbraham.
Warren M. H., Wilbraham.
Warren Miss E., Wilbraham.
Wilson Robert, Hartford.
Wright Mrs. S. F., Hartford.
Webster Mrs. Chas. F., Hartford.
Webster Chas., Hartford.
Wood Mrs., Upton.
Warner J. C., Granby.

Young Mrs. L. A. C., Warren.

.

www.ingramcontent.com/pod-product-compliance
Lightning Source LLC
Chambersburg PA
CBHW032358280326
41935CB00008B/622